"A seminal work in understanding the toll that killing remotely takes on the people who operate and support lethal remotely piloted aircraft (RPAs). It offers valuable context to leaders and technologists who seek to maximize automation and efficacy of RPAs while balancing the needs of the personnel who operate them. It also provides a reminder of the important role that humanity plays as our country moves to leverage big data and artificial intelligence for potential use in weaponized autonomous platforms. A must-have for the bookshelves of current and future leaders involved in lethal RPA or autonomous weapons programs."

—Major General James Poss (USAF, Ret.)

"An insightful look at a uniquely gifted community of warriors...Having served with Wayne Phelps on active duty, I am not surprised at all with the incisive analysis while capturing the human element of waging war remotely. Those who level criticism at the RPA and its dedicated service members must read this book, as should anyone who wants to know more about these dedicated warriors; time to give them the respect they deserve!"

—Lt. Gen. Robert F. Hedelund, USMC

"The psychology of RPA combat is as secretive as the airframes themselves—remote, mysterious, and hard to find. Wayne Phelps has taken the data, scientific and anecdotal, historic to present-day, and given it to us in an approachable, readable format. The veils of mystery of RPA combat are drawn back, such that one can understand the challenges, advantages, and adversity of this largely misunderstood type of warfare. Though the RPA enterprise is one of the 'most relevant and lethal' assets on the battlefield, it's often underrated and marginalized. This book gives much needed attention to the warriors who enter the virtual battlespace day after day, month after month, year after year in pursuit of the enemy, despite the clear and present danger of mental and moral injury. This book should be required reading not only for RPA commanders and aircrews, but for Intelligence troops, JTACs, TACPs, and all warriors in the ISR enterprise."

—Karen House, MA, MSW, LMSW, LPC, coauthor of *Avengers in Wrath: Moral Agency and Trauma Prevention for Remote Warriors*

"Phelps brings a crucial voice of analytical nuance to the discussion surrounding armed drones. Killing from a distance is not new, but the technical-political operational dynamic of drone strikes has generated a new mental paradigm for the war fighters our nation trusts to employ these weapon systems. *On Killing Remotely* will rank with *Our Robots, Ourselves* and *Reaper Force* in the canon of insightful and thought-provoking guides to how we consider the application of violence through technology."

—Ian MacLeod, former co-chair of the International Panel on the Regulation of Autonomous Weapons

"*On Killing Remotely* is an extraordinary achievement that provides a rich and intimate portrait of lethal Remotely Piloted Aircraft strikes and the operators who conduct them. Phelps illuminates how these operators are engaged in war in a novel and unprecedented way, which can have complex effects on how they see themselves as warriors and as moral beings. The book does this with sensitive attention to the organizational, cultural, psychological, and emotional dimensions of these operations, and how they combine in myriad ways to shape the experience of those involved in them. This work is destined to become a classic not only on the experience of warriors, or on war and technology, but on the fundamental quest of every human being to live a life with meaning."

—Mitt Regan, McDevitt Professor of Jurisprudence, Georgetown University Law Center; Senior Fellow, Stockdale Center for Ethical Leadership, United States Naval Academy

ON KILLING REMOTELY

ON KILLING REMOTELY

The Psychology of Killing with Drones

LIEUTENANT COLONEL WAYNE PHELPS
(USMC, RETIRED)

Little, Brown and Company
New York Boston London

Little, Brown and Company
Hachette Book Group
1290 Avenue of the Americas, New York, NY 10104
littlebrown.com

First Edition: May 2021

Little, Brown and Company is a division of Hachette Book Group, Inc. The Little, Brown name and logo are trademarks of Hachette Book Group, Inc.

The publisher is not responsible for websites (or their content) that are not owned by the publisher.

The Hachette Speakers Bureau provides a wide range of authors for speaking events. To find out more, go to hachettespeakersbureau.com or call (866) 376-6591.

ISBN 978-0-316-62829-7
Library of Congress Control Number: 2021936127

Printing 1, 2021

LSC-C

Printed in the United States of America

To the unsung heroes fighting our wars remotely.

Contents

CONTENTS

Introduction

A Predator Hunts for Bin Laden

> Man taxes his ingenuity to be able to kill without running the risk of being killed.
>
> —Ardant du Picq, *Battle Studies*

Two months after the unexpected and horrific terrorist attacks on America on 9/11, I was deployed to a remote location in Pakistan as part of a conventional force of Marines. By December 2001, the situation had significantly changed, as we were joined by another team that relocated from Uzbekistan to Pakistan. This team, known as a launch and recovery element (LRE), brought armed Predators with them. The team's arrival allowed me to bear witness to an event that would alter the fabric of warfighting: the first armed remotely piloted aircraft, flown over Pakistan and Afghanistan by Air Force pilots in the parking lot of the CIA headquarters in Langley, Virginia.

I played no part in the Predator's operations, but I became fascinated by the technology that enabled an aircrew to control an armed, remote-controlled airplane from the other side of the globe. I was enthralled by the very concept of it. The Predator was intriguing and the fact that it was armed was known only by a small circle of people at Big Safari, a U.S. Air Force test and rapid acquisition organization for special mission aircraft; General Atomics, the firm that developed the Predator and integrated the Hellfire missile; and the CIA.

The Predator program moved fast as a result of missed opportunities to kill Osama bin Laden. The rapid pace of the Predator's development was attributable partly to it belonging to an Advanced Capability Technology Demonstration, which meant that it was able to circumvent some of the traditional acquisition hurdles. It also helped that General John Jumper, the commander of the U.S. Air Force's Air Combat Command and eventually the Air Force's seventeenth chief of staff, was a proponent of the Predator and pushed for it to be armed.

The chief pilot of the Predator team was an Air Force special operations helicopter pilot named Captain Scott Swanson. Swanson first got involved with the Predator in 1998. Shortly after joining the program, he led the flight operations team and deployed the Predator to the Balkans. The Predator wasn't armed, but did include a newly integrated laser designator capable of marking targets for laser-guided munitions. After the Balkans, Swanson continued to serve as Big Safari's deputy for an Air Force team embedded with General Atomics Aeronautical Systems Inc. in San Diego.

In September 2000, an unarmed Predator launch and recovery element deployed to Uzbekistan to search for Osama bin Laden in Afghanistan as the U.S. grew increasingly concerned about the escalation of Al Qaeda attacks on U.S. interests. The LRE in Uzbekistan prepared the aircraft for flight, launched the airplane, and then "handed it off" to the mission control element (MCE). Handoffs were enabled through the use of satellite communications embedded in the front of the aircraft and communication and coordination between the LRE and MCE. The MCE, located at an American facility on Ramstein Air Base in Germany, unbeknownst to the Germans at the time, then commanded the flight. When the mission was complete, the mission control element handed the aircraft back off to the launch and recovery element for landing, refueling, and maintenance.

Swanson deployed to Ramstein as part of what Big Safari coined "the Summer Project." On two occasions during that deployment, Swanson spotted bin Laden in his camera. The first time was at an Al Qaeda training camp named Tarnak Farms. The crew of the unarmed Predator couldn't do anything except report the situation and wait for the appropriate authority to make the decision to launch a cruise missile or take other lethal action. Ultimately, cruise missiles didn't rain down on Tarnak Farms that day, and the opportunity passed. Bin Laden didn't resurface again for another decade.

The arrival of winter weather in 2000 prohibited the Predator from making the transit from Uzbekistan to Afghanistan over the Hindu Kush mountains and ended the Predator's first deployment to Afghanistan. The program accelerated over the next year. The first test shot of a Hellfire missile from the aircraft occurred at Indian Springs, Nevada, in February 2001. By September 2001, Big Safari and General Atomics had integrated the Hellfire missile and everything required to return to Afghanistan to hunt for bin Laden and Al Qaeda. And just in time.

On September 12, 2001, a day after the attacks on the U.S., an Air Force cargo jet, one of the only aircraft allowed to fly in the U.S. that day, departed the West Coast bound for the East Coast with Swanson, his team, and their equipment on board. Then Swanson and team set up in the parking lot of CIA headquarters at Langley, Virginia. On September 18, 2001, Swanson and his team resumed Predator flights over Afghanistan, this timed armed with Hellfire missiles. The LRE was simultaneously established at the previous launch and recovery location in Uzbekistan and remained there until relocating to Pakistan.

Three weeks later, on October 7, 2001, while I watched Tomahawk cruise missiles launch toward Afghanistan from the Arabian Sea before going ashore to Pakistan, Swanson, along with his sensor operator Jeff Guay, secured their footnote in history as the first Predator crew to fire a Hellfire missile in combat. Swanson recalls the details of that evening:

At one point, we tracked a couple of vehicles with Mullah Omar [the leader of the Taliban] heading out of Kandahar. Corroborating intelligence confirmed that it was Omar and that he was on the move. We ended up following him to a compound probably fifty to twenty kilometers outside of town where they stopped and went inside a building. We were on-site, on target there, preparing to assist inbound strikers to hit the target when a dialogue started on chat amongst our mission manager, the Air Operations Center [AOC], and intelligence personnel about whether to strike the target with fighters and whether a building in the area was a mosque.

They finally made a decision to take out a security team at the entrance of the compound in hopes that we could get Mullah Omar and his team to move to a different location where we could strike them. So, Jeff and I spun up the systems and a third person went over the checklist to make sure everything was OK. Our target was two guards in front of a Hilux pickup truck. We successfully launched a Hellfire and struck the target.

In the aftermath, it was interesting to observe the way the security forces on the ground reacted. They thought it was a ground attack coming from across the field and they responded like they were attempting to find out where it came from. The strike didn't achieve the result of flushing out Omar so the rest of the night there was infighting at the AOC about what to do next. They wouldn't hit the building that we knew Omar had gone into because we couldn't determine if the building directly next to it was a mosque. That opening night we had the ability to take out Mullah Omar and someone else made the decision we are not going to drop on the target.

Fifteen years after I first saw a Predator in Pakistan, I commanded an RPA squadron in the Marine Corps as a lieutenant colonel and deployed several teams to a foreign country to hunt down and kill more than one hundred violent extremists from the Islamic State. Our RPAs weren't armed with Hellfire missiles; instead they were unarmed and, upon finding enemy fighters, called in artillery strikes. It's more than a nuanced difference, and it led me to ask some basic questions about the human responses to killing across all the different categories of RPA.

By the time I left the Marine Corps, killing with RPAs had become common practice. Since killing permeates the Marine Corps' culture, I thought little about it at the time. It wasn't until I retired that I afforded myself an opportunity to reflect on what I had experienced and on what I had ordered others to do.

The inculcation into a killing culture, being surrounded by like-minded individuals, leads to glamorization and normalization of killing. I remember the first time someone pointed out to me the abnormality of this behavior outside of my culture. My wife and I approached the front gate of Marine Corps Base Quantico in Virginia and I was motioned to stop by a gate guard, a female lance corporal no older than twenty. I presented my military identification card to the Marine and she responded with a loud and resounding *"Kill, sir!,"* saluted, and waved me through the gate. I returned the salute, put away my ID, and drove on, thinking nothing more on the matter. My wife, a relatively new military spouse at the time, with a pronounced look of shock on her face, immediately asked, "Did she just yell 'kill'? Is that how you Marines greet each other?" My unapologetic answer was "yes." The reality is that the United States requires trained professionals skilled in the application of disciplined violence to enforce its foreign policy through military means. In my biased opinion, no organization nurtures that culture better than the United States Marine Corps.

Nearly two decades after the first Predator fired a missile in combat, unintended consequences and unforeseen repercussions of killing with RPAs have begun to emerge. One initially unforeseen issue is the negative psychological response to killing from maximum distance experienced by a percentage of RPA personnel. An Air Force study published in 2019 put the number of Air Force RPA personnel who experience post-traumatic stress disorder–like symptoms at 6.15 percent. At first blush, it seems counterintuitive that a person could be negatively affected by an action that occurs at a great remove, especially when that person isn't exposed to any physical danger. But over time, the notion that RPA personnel can suffer from PTSD gained acceptance among healthcare professionals. The question that remains is why does this happen? Some media postulate that RPA crews get PTSD as a result of a moral injury; where a person is forced to do something that differs from their moral beliefs.

In light of that landscape, I started this project with two simple questions: How does killing remotely with an RPA affect people and why does it affect them that way? To answer these questions I conducted my own research with the assistance of an Air Force MQ-9 Reaper pilot, Captain Justin Rhode. We surveyed 254 U.S. and foreign service members associated with the employment of RPAs, 245 of whom had killed with or through an RPA. Additionally, I conducted more than fifty in-person interviews to hear firsthand accounts of challenges and struggles. What follows are my findings on killing remotely.

Understanding Remotely Piloted Aircraft

Chapter One

The Evolution of Killing from a Distance

Humans are adept at finding mechanical means to overcome natural limitations. Humans were born without the ability to fly, so we found mechanisms that overcame this limitation and enabled flight. Humans were also born with constraints upon their ability to kill fellow humans, so throughout history we have devoted great effort to finding ways to overcome this resistance. From a combat evolution perspective, the history of warfare can be viewed as a series of successively more effective tactical and mechanical mechanisms to enable or force combatants to overcome their resistance to killing.

—Dave Grossman and Loren Christensen, *On Combat*

Warriors throughout history have continuously developed weapons that enabled them to overcome their physical and psychological limitations to kill. These tools ultimately made it physically and psychologically more effective to kill an enemy from greater and greater distances, moving the user farther and farther from danger. The culmination

3

of these tools, thus far in modern warfare, is an armed remotely piloted aircraft (RPA), colloquially and pejoratively referred to as a drone.

RPAs can be flown from the safety of one's home country by means of satellite communications, enabling the users to strike a target precisely, at the time of the user's choosing, from as much as seven thousand miles away, all while minimizing, mitigating, or eliminating collateral damage around the target. This evolution of killing from maximum distance has occurred over the span of human existence but increased exponentially in the twentieth and twenty-first centuries. Let's explore how weapons' ranges evolved from throwing a rock to launching a Hellfire missile off of an RPA on the other side of the Earth.

It Probably Began with a Rock

Although there is no documented history, it is easy to imagine that the first conflicts among early humans were resolved with bare hands, with the victor being the person who was strongest and could generate the most forceful blow. Since brute strength and overall size were not characteristic of every early man, that method undoubtedly gave way to the use of weaponry in resolving conflict: rocks, sticks, or clubs. Albeit primitive, these early weapons leveled the playing field as weaker beings could now overtake stronger ones by generating greater force with the aid of a tool.

As early man progressed, so, too, did the technology of tools. At some undetermined point humans discovered that a sharpened rock was more effective than a blunt rock, and sharpening rocks helped to concentrate and focus the kinetic energy, increasing lethality. (Although early man was probably not concerned with the level of lethality as much as he was with his own survival.)

Another discovery early man made was the ability to affix rocks to sticks to form a spear or weighted club. This advancement provided more standoff distance from an opponent and again increased kinetic energy through leverage.

Eventually early man figured out he could use a spear as a ranged weapon. In essence, this was the first true missile, since a spear could be launched at a distance. The first iconic missile was the next major weapon innovation: the bow and arrow.

Bow and Arrow: "The First Machine"

Exactly when the bow emerged as a weapon is unknown and debated among historians and archeologists, but rock-art scenes depicting archers from the prehistoric era were common finds in Africa and Europe. Due to the decaying nature of the materials from which early bows and arrows were fabricated, little archeological evidence exists. The earliest identifiable find was in Stellmoor, Germany, and dated around 9000 BCE, while the most definitive artifacts first identified as bows and arrows were found in Holmegaard, Denmark, and dated back to 6000 BCE.

Regardless of when bows and arrows came into existence, the impact this weapon had on early humans' ability to hunt and wage war secures its place in history as one of the most important early inventions. John Keegan in *A History of Warfare* described the bow as the first machine, spreading rapidly once invented. Keegan continued that after the bow, "man no longer had to close to arm's length to dispatch his prey, pitting at the last moment flesh against flesh, life against life. Henceforth, he could kill at a distance."

The implication being that killing at a distance was obviously preferred. After the introduction of the bow, warriors had the ability to enjoy some standoff distance and relative safety while employing a weapon that could kill their enemy.

Around 2200 BCE, evidence appeared that the Sumerians developed a composite bow that doubled the range of the simple bow and increased its penetrating power. Richard Gabriel and Karen Metz described the introduction of the composite bow in a report for the Strategic Studies

Institute of the U.S. Army War College titled "A Short History of War: The Evolution of Warfare and Weapons":

> This bow was a major military innovation. While the simple bow could kill at ranges from 50–100 yards, it would not penetrate even simple armor at these ranges. The composite bow, with a pull of 2–3 times that of the simple bow, would easily have penetrated leather armor, and perhaps even the early prototypes of bronze armor that were emerging at this time. Even in the hands of untrained conscript archers, the composite bow could bring the enemy under a hail of arrows from twice the distance as the simple bow. So important was this new weapon that it became a basic implement of war in all armies of the region for the next fifteen hundred years.

Pairing the composite bow with the chariot provided a military advantage in terms of mobility, maneuverability, and range that made for a dominant tool lasting from 2000 BCE to the first century CE. Opposing armies eventually learned to counter chariot archers by targeting the horses pulling the chariot, by maneuvering into rough terrain unnavigable by chariots, by fighting in formations, and by using arrow-proof shields.

The Chalcolithic period, defined as the period where copper and stone coexisted, succumbed to the Bronze Age. Keegan described this transformation: "man bent to the almost universal rule that a superior technology obliterates an inferior one as fast as the necessary techniques and materials can be acquired." Weapons became more sophisticated as bronze could be fashioned into larger blades than stone, resulting in edged weapons such as swords, knives, and axes. While these advancements had significant impact on the formation of professionalized and well-armed armies, they did not increase the range at which one could kill their enemy.

Bronze Age weapons were supplanted by Iron Age weapons once the ability to create weapons from iron ore became widespread. Weapons of the Iron Age were forged rather than casted, permitting a smith to manipulate the material to form and keep a sharp edge on a weapon, particularly a sword. Iron Age weapons proliferated on a larger scale than Bronze Age weapons, mainly because of the abundance of iron. This allowed for mass production of weapons and the arming of even larger fighting forces. While this enabled war on a larger scale, the weapons of choice were still edged weapons and the standoff range between combatants did not increase.

The bow and arrow continued to dominate this era as the most popular standoff weapon. It was such an effective weapon that its use continued through many variations and improvements until eventually it could not compete with the effectiveness, speed of employment, and range of firearms.

Gunpowder: Everything Changes

The invention of gunpowder by the Chinese in the ninth century CE propelled the next advancement in killing from a greater distance, although the range advantage was not immediate. In fact, early firearms were still outranged by crossbows and longbows. As discussed in *On Combat*:

> The longbow and the crossbow were significantly more accurate, with a far greater rate of fire and much greater accuracy range than muzzle-loading muskets used up to the early part of the American Civil War. Furthermore, the longbow did not need the industrial base (iron and gunpowder) that muskets required. [...]
>
> Mechanically speaking, there are a few reasons why there should not have been regiments of longbows and crossbows at Waterloo and Bull Run, cutting vast swathes through the enemy.

The advantage of early firearms was psychological in nature more than physical, producing a loud sound and an accompanying puff of smoke that filled the battlefield and served to overwhelm the senses and frighten the enemy into running, and it was in this retreat that the vast majority of killing occurred. On a battlefield where will and nerve greatly contributed to an army's fighting ability, courage, and success, early firearms provided an advantage not experienced with longbows and crossbows.

This idea of posturing is not only human nature but also visible in the animal kingdom. A dog will raise its hackles, growl, and bark loudly when confronted by another potentially threatening dog. A cat will arch its back and hiss in an attempt to appear larger and more threatening. Peacocks flare out their feathers to display a multitude of colors and appear larger. And humans will yell, square up, extend their arms to appear bigger, and attempt to stare down an opponent in a threatening manner.

A prime example of a human ritualistic practice to intimidate other humans is the haka. The haka was a pre-battle war dance originally performed by Maori warriors designed simultaneously to intimidate the opponent and to prepare the Maori warrior psychologically to conduct violence. Any observer of the All Blacks from New Zealand performing this ritual prior to a rugby match knows the immediate psychological advantage this posturing gives the participant. Early guns produced a similar effect, bolstering the confidence of the user.

The true range advantage from gunpowder came in the form of cannons. Cannons were the next logical progression in the artillery family. Prior to the cannon, artillery consisted of catapults that hurled stones, or large versions of crossbows that fired bolts. Both weapons relied on stored mechanical energy to propel an object. According to Gabriel and Metz, Philip of Macedon was the first to designate a unique group of artillery engineers to design catapults as siege weapons.

Alexander the Great took that one step further and used artillery as cover for infantry. Alexander's army marched with preconstructed catapults among its ranks and pulled larger disassembled catapults in wagons that were assembled prior to battle. The Romans continued to improve upon early Greek artillery and developed a large torsion device called an onager (or wild ass), capable of hurling a hundred-pound rock more than four hundred yards. The Romans integrated artillery into all of their battle formations and were the first to use it effectively as an antipersonnel weapon.

When cannons appeared on the battlefield, they replaced the preexisting artillery in the roles of siege and antipersonnel weapons as well as yielding greater strategic value over artillery. Cannons varied in range and accuracy based on their size and on the proficiency of the crew, but they were the longest-ranging weapons of their time. Their mere presence on the battlefield produced a posturing effect.

The introduction of the limber (a two-wheeled horse-drawn cart holding the cannon barrel) near the end of the fifteenth century helped the cannon proliferate throughout Europe and Asia. This newfound mobility ensured armies could transport their firepower to the battlefield. Most important, since cannons were manned by a crew and fired at an enemy at a range greater than that of other existing weapons, they helped overcome the physical and psychological limitations of killing. The cannon and its successor, field artillery, remain among the most prolific killing weapons in history.

Bessemer, Minié, and the "Second Industrial Revolution"

Iron cannons continued to be the dominant range weapon on the battlefield until the second industrial revolution and the mass production of steel in the mid-nineteenth century. The British engineer Henry Bessemer patented a process in 1856 bearing his name that revolutionized the steel industry.

The genesis of Bessemer's arrival at this point is an interesting story rooted in an effort to design rifled projectiles for smooth-bore cannons during the Crimean War. According to Bessemer's autobiography, "It was under the impression that my invention would enable all existing smooth-bore guns to be at once utilized for discharging elongated shells and solid projectiles, and would at the same time solve a problem of great national importance, that I applied for and obtained a patent on the 24th November, 1854."

His proposed invention was submitted to the British War Department for consideration and was summarily rejected. Intent on proving its viability, Bessemer designed a small mortar to fire the projectile and tested it on his own land. The initial experiments were successful, demonstrating the desired rotations of the projectile in flight.

A few months later, Bessemer attended a dinner party in Paris, where he was introduced to Prince Napoleon, Emperor Napoleon III's cousin.

Bessemer's rifled ordnance was discussed and the prince invited Bessemer to pitch the idea to the emperor.

Napoleon III, whose country was also engaged in the Crimean War, liked the idea and funded Bessemer to refine the invention. Bessemer developed the projectiles and a suitable cannon and demonstrated them in France to an audience of French generals, particularly Claude-Étienne Minié, whom the emperor sent to observe and report the results of the experiment. The remainder of the story as told by Bessemer in his autobiography:

> Commandant Minié remarked that it was quite true that the shot revolved with sufficient rapidity, and went point forward through the targets; and that, he said, was very satisfactory as far as it went. But he entirely mistrusted their present guns, and he did not consider it safe in practice to fire a 30-lb. shot from a 12-pounder cast-iron gun. The real question, he said, was; Could any guns be made to stand such heavy projectiles? This simple observation was the spark which has kindled one of the greatest industrial revolutions that the present century has to record, for it instantly forced on my attention the real difficulty of the situation, viz.: How were we to make a gun that would be strong enough to throw with safety these heavy elongated projectiles?

The answer to Bessemer's question was steel, and the process he developed to manufacture it cheaply in large quantities accelerated the second industrial revolution.

In addition to steel, spiraled grooves, called rifling, inside musket and cannon bores gained prominence around this time. Rifled bores forced projectiles to spin, which improved the accuracy and range of both weapons. A rifled musket could be effectively fired out to a distance of 180 yards, farther than smooth-bored muskets by a factor of three. By the end of the Civil War, a rifled musket combined with a newly shaped bullet that formed a tighter fit in the bore's rifling (called a "Minié ball" and named after its inventor, the same Commandant Minié who inspired Bessemer) enabled soldiers to kill targets out to a range of one thousand yards.

Keegan noted in *A History of Warfare* that Alfred Krupp, a German steelmaker with a family history of making weapons, first started casting steel cannons around 1850. Krupp perfected this innovation over the next decade through the Bessemer process and open-hearth steel making. By the end of the nineteenth century, Krupp's guns (ranging in caliber from 77 to 155 millimeters) had a reputation for high quality and proliferated throughout forty-six countries.

The Crimean War from 1853 to 1856 witnessed the first use of breech-loading rifled cannons by the British Army, and by the end of the Civil War in 1865 half of all Union cannons adopted this new technology, which offered greater accuracy and range. Advancements in artillery range improved to the point that by the time of the Russo-Japanese War in 1904–05, forward observers armed with field phones were needed to relay back to the gun-lines the location of targets and to direct fire. Humans' ability to kill from a distance had officially moved beyond the visual range of those who employed the weapon.

The Krupp industry continued to advance the range and size of artillery pieces, introducing two 420-millimeter cannons prior to World War I. These monstrous cannons were nicknamed "Big Bertha," most likely after Bertha Krupp, granddaughter of Alfred, who owned the Krupp industry at the time. Big Bertha was considered a siege weapon and was effectively employed out to a range of nine kilometers. In World War I, two Big Berthas were sent to the Battle of Liège and were successfully used to destroy two Belgian forts. During the interwar years, Krupp developed two 800-millimeter railway guns, aptly named for their attachment to railway cars, that could fire a seven-ton projectile more than thirty-seven kilometers. It was the equivalent of launching an elephant just short of the length of a marathon.

Another important by-product of the second industrial revolution was the mass production of arms and ammunition. Advancements in weaponry and manufacturing matured by the end of the nineteenth century and were prominent as the Great War erupted in 1914. One such example was the machine gun, developed in 1884 by the American Hiram Maxim. Maxim built an automatic weapon that operated like a machine, firing up to six hundred bullets a minute, where the machine's mechanical function necessary to fire the next bullet was powered by the detonation of the previous bullet. These machine guns fired out to a range of a thousand yards, were organic to infantry units, and were manned by a crew.

The machine gun helped its users overcome several physical and psychological limitations to killing. It increased the effective range, speed, rate of fire, and volume of ammunition discharged. The machine gun exponentially increased the killing power of the gun crew since it fired the same volume of bullets in an equivalent time as one hundred soldiers armed with rifles. It also helped to overcome psychological limitations to killing as a crew-served weapon. *On Combat* makes the point: "a soldier can make the decision in his own heart and soul not to pull the trigger, but if he is part of a team, such as a crew-served weapon, he would have to talk to the other soldier about it and they would have to agree not to shoot. This rarely happens."

Air Warfare: "Slipping the Surly Bonds"

World War I was dominated by machine guns and artillery. However, the war saw the introduction of an interesting machine that enabled humans, according to the poet John Gillespie Magee Jr. in his sonnet "High Flight," to "slip the surly bonds of earth and touch the face of God." Humans first left the ground in powered flight on December 17, 1903, in Kitty Hawk, North Carolina, less than eleven years prior to the outbreak of war in Europe. The airplane revolutionized warfare, but not in its nascent stage during World War I, where its impact on the overall outcome of the war remains debatable.

At first, airplanes were used as a reconnaissance platform, which they were ideally suited for. Shortly thereafter, airplanes were transformed into aerial fighters with the addition of a machine gun or served as rudimentary bombers. By 1915, the combination of the machine gun and the airplane had a similar effect as the chariot and archer combination: unparalleled mobility and lethality. Airplanes added dimension and range to war.

Although airplanes' range during World War I was insignificant compared to modern aircraft, they afforded the pilot the ability to take off from the relative safety of an aerodrome, engage in aerial combat, and return to the place of origin, where the pilot laid his head on a pillow at night without the interruption of constant shelling from artillery. This is the first major step in truly "slipping the surly bonds" of earthly mud and grime and escaping the constant threat of death on the modern battlefield, as we study the evolution of killing remotely.

During the interwar years pioneers accomplished feats that significantly advanced aviation. The first solo transatlantic flight from New York to Paris was conducted by Charles Lindbergh on May 20, 1927. This grueling 3,600-mile flight took 33.5 hours but made Lindbergh a household name worldwide overnight. From a military perspective, this achievement demonstrated the possibilities of long-range aviation operations that emerged in World War II. James "Jimmy" Doolittle's famed "blind flight" on September 24, 1929, paved the way for the use of navigational instruments.

According to Winston Groom in *The Aviators: Eddie Rickenbacker, Jimmy Doolittle, Charles Lindbergh, and the Epic Age of Flight*, "By this singular, decisive feat Doolittle advanced aviation into the modern age, in which weather would no longer be a controlling factor in flying." Doolittle once again pushed the boundaries of the possible following the attacks on Pearl Harbor, when he led a one-way bombing raid of sixteen B-24s launched from an aircraft carrier toward the Japanese capital of Tokyo. While not tactically significant, strategically the Doolittle Raid proved that America could strike into the heart of the Japanese homeland and served to shore up American will and morale.

The advancement of weapons during World War II proved what humans were capable of accomplishing when the world's largest industrial complexes focused on inventing and manufacturing weapons that provided tactical and strategic advantages for their nations. Firearms, artillery, tanks, vehicles, airplanes and ships were mass produced on an unprecedented scale as the world's powers fought a total war of attrition. Although the numbers and capabilities of military aircraft swelled and their combat radius increased significantly during World War II, the distance between an airplane and the target on which it could drop a bomb improved primarily in altitude rather than in standoff distance from the target.

Even the heaviest bomber of World War II, the B-29 Superfortress, with a combat radius of two thousand miles, carrying twelve thousand pounds of bombs and twelve .50 caliber machine guns for self-defense, still flew directly over the target area to drop its bombs, and in order to employ its self-defense armament needed to be within machine-gun range of attacking enemy fighters. The advancements of aircraft in World War II included increases in combat radius, altitude, speed, and payload capacity. But the ordnance employed from those World War II– era aircraft still required a large degree of exposure to danger for the crew. Killing remotely—the combination of killing from maximum range while removing the combatant further from danger—had still not been achieved for aircraft by war's end.

World War II also witnessed the development and use of the first nuclear weapons. The destructive capacity relative to the size of the weapon was unlike anything ever seen before when a B-29 nicknamed the *Enola Gay* dropped Little Boy on Hiroshima on August 6, 1945, killing seventy thousand people nearly instantly. But even Little Boy and Fat Man, dropped on Nagasaki three days later, required carrier platforms full of humans to fly over the target area. This harnessing of a nuclear reaction to form a military weapon kicked off an arms race and started a cold war among the world's superpowers, but nuclear weapons' full killing potential weren't realized until coupled with another technology progressed by the Germans during the war: rockets.

Rockets: From V-1 to SCUD to ICBM

Rockets had existed in various lesser forms since the Chinese invented gunpowder in the ninth century. But it was the Germans who advanced the next stage of killing from a remote distance with the invention of the V-1 and V-2 guided missiles. Keegan in *A History of Warfare* described Hitler's insistence on developing these weapons:

> The manned bomber was a fragile weapon of offence, as Hitler had learned to his cost in the campaign of 1940 against Britain. That was the principal reason for his enthusiastic espousal of a programme of unmanned aircraft development generously funded by the army since 1937. In October 1942 a test firing of a rocket with a range of 160 miles, designed to carry a ton of high explosives, had taken place, and in July 1943 Hitler declared it "the decisive weapon of the war" and decreed that "whatever labour or materials (the designers) need must be supplied instantly."

Hitler referred here to the V-2 rocket, which was first employed very late in the war, in September 1944, flew in an unguided arching trajectory, and was recorded as the first human-made object to enter into space. Hitler's prediction that the V-2 would be the decisive weapon of the war never came true. In the end, the Germans fired only 2,600 V-2s against Britain, and although the V-2 was a terrifying psychological weapon, its tactical significance was minuscule.

The V-1 was the precursor to a modern-day cruise missile. It was an unmanned, guided missile that flew a predetermined path out to a range of almost 160 miles. V-1s were first launched in June 1944, and at their apex of employment more than a hundred a day were fired toward London. V-1 and V-2 rockets were obviously employed beyond visual range, but what they gained in range they lacked in accuracy and were no more discerning of targets than Saddam Hussein's SCUD missiles fired toward Israel during Operation Desert Storm. Regardless, the foundation was built for future rocket technology that landed in the hands of the United States and the Soviet Union after World War II.

Rockets paved the way for weapons to increase their range more than ever before. The V-2 concept of a guided ballistic missile produced an arms race between the United States and the Soviet Union during the Cold War that might best be described as mutually assured destruction from maximum range. Intercontinental ballistic missiles (ICBMs) armed with nuclear warheads provided the ultimate standoff range with maximum destructive power.

The main deterrent to employing a nuclear-tipped missile was the threat of a returning volley of missiles being fired on one's homeland. This Cold War played out in the space race as well, as the U.S. and Soviet Union fought for primacy in that new frontier. The space race gave rise to miniaturization of components and increased onboard computing power, which enabled development of tactically employed missiles ranging from shoulder-fired antiaircraft missiles to cruise missiles.

Further by-products of the space race were satellite communications and the global positioning system (GPS), which together enabled precise position reporting, navigation, and strike. The inclusion of lasers, electronic targeting optics, and GPS-guided munitions further increased the range at which tactical platforms could engage a target.

Bombs, Dumb and Smart

In the Vietnam War, U.S. forces attempted to destroy a bridge in 1965 during Operation Rolling Thunder. The bridge, nicknamed the Dragon's Jaw due to the tenacious defenses surrounding it as well as its sturdy construction and seeming invulnerability to "dumb bombs," crossed the Song Ma river just seventy miles south of Hanoi. The bridge was used as a major supply route for the North Vietnamese, and it was such a strategic target that it was number fourteen on the U.S.'s original list of targets to destroy at the opening of the war.

Seven years later, the bridge was still standing despite having been hit multiple times. In the process, the U.S. had sent a total of 871 sorties to attack the bridge and had lost eleven aircraft. But that changed with the introduction of "smart munitions." On May 13, 1972, F-4 jets armed with twenty-six laser-guided bombs attacked the Dragon's Jaw and finally damaged the bridge enough that it would take several months to repair. Smart weapons changed the dynamic of targeting. Whereas it previously had taken several sorties to destroy a target, with the introduction of smart munitions multiple targets could be destroyed in one sortie, all from a greater standoff range from the target and with much greater accuracy.

Cowardly or Chivalrous? Bows, Guns, Aircraft, and RPA

At the turn of the second millennium, several technologies converged into a platform that enabled a crew to conduct, at the time of the crew's choosing, a precise tactical strike on a target seven thousand miles away. Warriors completely removed from harm's way could now kill an enemy combatant from maximum remote distance and be home in time for dinner with their family. While at face value this sounds like the apex of removing all physical and psychological limitations to killing, there are always unintended and initially unrecognized challenges with any new weapon.

The evolution of weaponry has obviously demonstrated a trend toward increasing distance between combatants. As every instance of a major advancement in killing from greater distance occurs, society decides how and whether it accepts this new form of killing and whether it coincides with the generally accepted concept of a warrior's code. This decision tends to dictate the initial attitude toward the weapon (and those employing it) until it gains greater prominence and ultimate acceptance by society.

One might think that an effective killing weapon such as the bow and arrow would have been widely accepted by society and that other weapons requiring closer distance (such as sword fighting) would have been immediately abandoned. However, every weapon has its limitations and purpose and the bow and arrow was no exception. A bow required its user to be physically conditioned and trained well enough to be accurate; otherwise, it was no factor on the enemy. An archer's quiver could hold only a couple dozen arrows, which presented the logistical challenge of keeping the archers supplied. This also left open the question of what to do with archers in battle after their supply of arrows was exhausted. To be an effective archer demanded immense training and time, and a leader wouldn't want to risk losing archers to a sword fight in which they were ill prepared to compete.

There were also societal challenges to accepting this new form of killing from a distance. In England, in the late twelfth and early thirteenth centuries, the bow was viewed as an unchivalrous peasant's weapon. To duel face-to-face, with both opponents equally invested in the fight and the possibility of death, was considered an honorable way to fight. Killing from a distance contrasted with the social code of honor and courage practiced by the knights of the day. But there was no denying the effectiveness and widespread use of the bow, and in time it was accepted by English nobility as a necessary part of warfare. This was evident at the Battle of Crécy in 1346, when Englishmen armed with long bows defeated French mounted knights.

Much as archers in the thirteenth century faced social scrutiny for fighting with a weapon initially considered cowardly or dishonorable, employers of early firearms also experienced initial apprehension in the sixteenth century. John Keegan brilliantly describes this in *A History of Warfare*:

> Though they had admitted gunpowder technology to their traditional practices, they had not adjusted to its logic. Like the Mamelukes who bore down, sword in hand, on the firearms of the Egyptian sultan's black slaves, they were still trapped in an ethos which accorded warrior status only to horsemen and to infantry prepared to stand and fight with edged weapons. Fighting at a distance with missiles was beneath the descendants of the armoured men-at-arms who had dominated European warmaking since the age of Charlemagne.

The introduction of the airplane in warfare was not met with the cultural and societal aversion that the bow and firearms had previously experienced. Rather the opposite. Why pilots and airplanes experienced this level of support, and why other revolutionary weapons weren't initially so well received, is unclear, but the standard of comparison seems to be similar to the knight's moral code of fighting. Whereas bows and guns were at first viewed as cowardly weapons, airplanes harkened back to the days when knights faced their opponent in open battle, with both opponents physically and equally invested in danger. In fact, the first fighter pilots were often referred to as the knights of the air.

Early pilots' courage was lauded, their kills were viewed as just and moral, and their stories were inspiring tales used to rally patriotic support for the war effort. Perhaps this reputation was earned by the courage required even to fly this new, dangerous, and frequently unreliable weapon in its infancy. Perhaps it was the nature of the aerial duels of single combat won through skill, mastery of machinery, and bravery. Or perhaps it was because the ground forces were gridlocked in the trenches fighting in an unglamorous manner, and everyone longed for a hero.

Equally shared danger and mutual respect for one's foe categorized fighter pilots of World War I. The German pilot Manfred von Richthofen, better known as the Red Baron, was credited with eighty air-to-air kills. As anecdotal evidence of the respect accorded to opposing knights of the air, when the Red Baron was shot down, the British buried him with full military honors. Regardless of the societal reasons for acceptance of the airplane from the outset, this was the petri dish that the military aviation culture originated and grew in and that, less than a hundred years later, remotely piloted aircraft threatened to alter, despite the fact that unmanned aircraft actually preceded manned aircraft in flight.

On May 6, 1896, seven years before the Wright Brothers took flight at Kitty Hawk, Samuel Langley launched a steam-driven unmanned aircraft over the Potomac River near Quantico, Virginia. Langley's "Aerodrome Number 5" flew for ninety seconds over half a mile at a speed of twenty-five miles per hour at an altitude of eighty feet above the water. By November 1896, Langley flew "Aerodrome Number 6" almost a mile. For each flight, Langley used a houseboat with a catapult to launch the aircraft into the air.

Theodore Roosevelt, who was assistant secretary of the U.S. Navy at the time, saw potential in this machine and recommended that the Navy study uses for it. The Navy did see utility in the aircraft; however, it passed on sponsoring Langley with funding. The U.S. Department of War seized on the idea and sponsored the initiative with $50,000 in grants and one caveat: Langley was to build a version capable of carrying a person. Langley's first attempt was on October 7, 1903, and resulted in an immediate crash into the Potomac. Langley's second attempt, on December 8, 1903, met the same fateful result. Nine days later, the Wright Brothers successfully flew at Kitty Hawk, thus changing the course of aviation and burying unmanned aviation's origins in the inconsequential footnotes of history. Until the turn of the second millennium, when technology converged to allow for an armed remotely piloted aircraft to be flown from the opposite side of the globe.

Air Force Colonel Joseph Campo described the context of how RPAs' introduction fits with the historical struggle to accept new weapons in his article "Distance in War: The Experience of MQ-1 and MQ-9 Aircrew":

> This grander context suggests that MQ-1/9 simply represent another step in the evolution of distance-based warfare and not the crossing of some imaginary precipice that we should attempt to turn back from. Based upon centuries of military technological and cultural advancements, we should have anticipated the cries of airmen, soldiers, and statesmen lamenting the morphing of warfare via MQ-1/9 into a video game devoid of seriousness and lacking any recognition of the noble warrior traditions currently in use. They echo the slow acceptance rates of previous weapons and methods for much the same rationale. Bowmen, riflemen, and snipers all experienced similar critiques as they were introduced onto the battlefield and grudgingly accepted. MQ-1/9 aircrew have fared no better or worse in this regard. The cycle of critique-accept-repeat is a trend that has persisted for centuries.

Based on perceptions and portrayals of RPA in the media, literature, the entertainment industry, international opinion, and how RPA personnel are treated within the military, it's safe to say that we are still in the critique stage of this weapon system. Unfortunately, this has immediate impact on those employing the new technology, as we will discuss in the following chapters.

Chapter Two

An Insatiable Appetite: Rise of the Robots

Before the war, the Predator had skeptics, because it did not fit the old ways. Now it is clear the military does not have enough unmanned vehicles. We're entering an era in which unmanned vehicles of all kinds will take on greater importance—in space, on land, in the air, and at sea.

—George W. Bush, forty-third president of the United States, in an address to the Corps of Cadets at the Citadel on December 12, 2001

A Commander's Appetite

Around the turn of the millennium, killing from maximum distance was enabled by the convergence of technologies into one platform. But it was military commanders' insatiable appetite for persistent intelligence, surveillance, and reconnaissance (ISR) and full-motion video in operations centers during two simultaneous wars that fueled the rapid expansion and use of remotely piloted aircraft in the military. The military had certainly dabbled in RPA in the past, but early models were unreliable. Following the development and employment of an armed Predator, the conditions were ripe for the RPA atmosphere to change. All that was needed was a catalyst.

The terrorist attacks on the United States on 9/11 and the subsequent wars in Afghanistan and Iraq were the catalyzing events that changed the RPA landscape. Two prolonged counterinsurgency fights signaled the immediate and unquenchable demand in the U.S. military for an "unblinking eye in the sky" with one initial primary purpose: to watch over troops. Justifiably so, as the biggest threat to U.S. forces during those wars were undetected improvised explosive devices (IEDs) placed alongside roads and employed with devastating lethal effects.

Fighting two simultaneous counterinsurgency wars meant that the enemy wore plain clothes, hid among the civilian population, and used guerrilla-style fighting tactics. Locating that enemy required massive amounts of persistent surveillance that did not exist prior to the advent of RPA. There were never enough troops deployed to prevent the enemy from emplacing IEDs along the roads or trails. Nor was there ever a shortage of artillery shells used for IEDs in Iraq, homemade explosives in Afghanistan, or insurgents willing to emplace IEDs in either country. Besides, guarding a road is a very mundane task for a soldier and not a good use of limited resources, especially personnel. A solution other than a soldier occupying terrain was required.

Constantly watching a road was a great mission for a machine that specialized in dull, dirty, and dangerous work. RPAs filled the tactical and logistical gaps amplified in counterinsurgent warfare and turned these deficits into advantages in a manner no technological advancement of warfare was previously able to do.

RPAs also provided an unprecedented tactical and often strategic advantage: the streaming of real-time imagery and video. Video streams from Predators, Pioneers, Scan Eagles, and the like were displayed in military operations centers on front lines, the Pentagon, and echelons in between in the U.S. and overseas. Commanders at every level watched the battle unfold before their eyes in real time. Video feeds from the front lines offered access to and control of information that heretofore never existed in operations centers, a frontline perspective that shifted on demand with a simple chat message to an RPA sensor operator. Live video feeds and instant reach-back to aircrews helped commanders make decisions. The information gleaned from video feeds turned uncertainty to certainty, built situational awareness, sped up a commander's decision-making process, reduced risk, and enabled commanders to control the tempo of the fight.

Of all the advancements to the modern war machine, it was the live-motion videos of the battlefield, streamed from RPA, that captured the most attention. At any given time during the wars in Iraq and Afghanistan, multiple video monitors in an operations center displayed RPA tactical video feeds of the battlefield; a firefight here, an ambush there, and an IED explosion on a third monitor. Watching video feeds had an effect on the viewer that can best be explained by a phrase commonly used at the time: "Predator porn." The effect on viewers was similar to that of a flame on a moth. It was inviting. It drew you in. It was combat voyeurism at its best. Those rare glimpses of catching some action on a video screen made the viewer feel more connected to the war they participated in, despite the fact that their work in an operations center was far removed from the tactical fight.

Fortunately, RPA video feeds served more of a tactical purpose than just background video noise in an operations center. The feeds helped protect frontline troops on the battlefield. Countless hours were spent observing "patterns of life" to determine a predictable baseline of normal and abnormal activity of the indigenous populations of host nations. Route reconnaissance missions swept roads to look for IED trigger men, indications of IEDs buried on the side of the road, or enemy ambushes lying in wait. Diligent effort, combined with the ever-present goal of reducing U.S. and allied casualties, turned the voyeurism of Predator porn into a worthy scholarly endeavor.

Furthermore, RPA video feeds played an important role in the conduct of air and artillery strikes. By the end of Operation Iraqi Freedom in 2011, RPA video surveillance was so critical for building awareness of a target that Joint Terminal Attack Controllers, the observers who call in air or artillery strikes, wouldn't strike a target without a video feed. While the video wasn't actually required to strike a target, it was definitely perceived as an enhancement tool that could help minimize collateral damage, shorten the time to cue striking aircraft onto the target, and assess the effectiveness of the strike.

The demand for full-motion video at all levels on the battlefield resulted in increased numbers of RPAs requested to support operations. The requirement emerged so rapidly that it quickly outpaced what little organic assets existed within the U.S. Department of Defense. But the defense industrial complex was happy to meet military need by leasing video time from their own unmanned aircraft while the DoD attempted to ramp up.

Other technological advances that emerged at the same time as RPAs also helped to enable their rapid expansion. At the onset of the wars in Afghanistan and Iraq, large-scale chat rooms enabled a flattening of the traditional military command and control hierarchy. Historically, military orders were passed from commander to subordinate commanders in person, via written orders, or over the radio. Orders were passed from subordinate commanders and so on down the line until the orders reached frontline troops for execution. As Thomas Friedman discussed in *The World Is Flat,* the internet connected the globe and rapidly expanded around the turn of the millennium, consequentially flattening traditional structures and streamlining once cumbersome processes.

The same flattening of traditional hierarchies occurred in the military command and control architecture. Digital communications such as email and chat rooms supplanted traditional methods of communication. Commanders monitored secure chat rooms where critical time-sensitive warfighting information was passed and shared rapidly and simultaneously with everyone in the chat room. A commander could even chat directly with the crew of an RPA if desired, which we'll discuss later in the chapter "Demands of Authority."

The integration of new technology in an operations center provided commanders with the ability to track the precise location of subordinate forces, digitally communicate with anyone they desired, and watch the battle unfold on a live-motion video feed. It was an unprecedented fusing of information in a single tactical hub that was met with an unprecedented appetite for more.

The Birth of a Predator

Despite the burgeoning demand, the use, proliferation, and now necessity of RPAs in combat was not always planned or even anticipated. Perhaps the first indication that RPAs were more than just a novelty item was the battle of Roberts Ridge in Afghanistan during Operation Anaconda. In "Phantom of Takur Ghar: The Predator above Roberts Ridge," Senior Airman James Thompson recorded the details of events of one fateful day:

> A once quiet landscape turned battlefield, the clash of gunfire and shouts ripped through the Shahi-Kot Valley in the early hours of March 4, 2002. As part of an early war effort that targeted al Qaeda and Taliban forces in Afghanistan, the Battle of Roberts Ridge is still known as one of the deadliest engagements during Operation Anaconda.
>
> Above the Takur Ghar mountain top, an MQ-1 Predator aircrew became an unforeseen, close air support asset for a desperate joint special operations team in their time of need.
>
> Deep, black smoke from a crashed, bullet-riddled MH-47 Chinook helicopter filled the air. Among the wreckage were wounded, the lead combat controller on the ground, Maj. Gabe Brown, then a staff sergeant, along with the rest of the special operations team who worked to secure casualties and defend their position on the summit.
>
> Pinned down on the landing zone and under direct fire, Brown established communications with an MQ-1 aircrew in the area who had visual of the team. Col. Stephen Jones, then captain and Predator pilot, had already been in the cockpit and was ordered to support just moments after the crash.
>
> Before Jones arrived on station that early morning, he had no idea what he and his team were in for.

"I remember coming in on shift that night and there was a lot of commotion," Jones said. "I was told to get out to the ground control station as soon as possible."

Throughout the day, Brown said he developed rapport with the Predator pilot as he gave situational awareness updates and assisted with targeting enemy combatants.

"When I had fighters check in, he would buddy lase [mark the targets with Predator's laser] for those inbound fighters and would help me with the talk-on, so it cut my workload dramatically having him there," Brown said.

Many other U.S. and coalition aircraft were simultaneously entering and exiting the area. Before authorizing a strike, Brown needed to "talk-on" the respective aircrew, which meant he briefed the situation on the ground to every aircraft that entered the airspace.

With a bird's-eye view, Jones and his aircrew alleviated some of Brown's duties and took control of liaising information within the zone, while serving as forward air controllers in the battle.

"(From our cockpits) we were serving as forward air controllers airborne or FACA, and I was serving as the on-scene commander," Jones said.

He began looking after the survivors, deconflicting airspace for coalition aircraft coming in and out, as well as communicating back to the joint command and control elements about the survivors' condition as they put together an evacuation plan.

"Gabe was doing a phenomenal job being a controller on the ground calling in close air support, but it was a lot of work," Jones said. "There were a ton of coalition aircraft coming in and out and some of them didn't have much play time, meaning they had to get in, develop an understanding of what was going on, receive a nine-line and then drop bombs or shoot their missiles."

The aircrew took some of the burden from Brown who remained on frequency with Jones, ready to voice commands at any moment.

Brown was able to conserve radio battery life due to the aircrew's initiative and the MQ-1's ability to loiter over the battlefield for extended periods of time.

Ground forces were still pinned down from continuous bunker fire when Jones relayed the evacuation plan to Brown. Their team was in need of a precise air strike that could eliminate the enemy hunkered down deep in the mountainous terrain.

Brown first called upon fighter aircraft.

"We were basically trying to use walk-in ordinance off the fighters, using 500-pound bombs to frag (blast) the enemy out of the bunker and we were unable," Brown said.

After numerous attempts, Brown and his team were running out of options and daybreak quickly approached...

Brown and his team were considered danger-close due to their proximity to the target, causing concern for aircrew and senior leaders. However, Brown's need for immediate aerial support outweighed any apprehension.

"It was late in the morning, he (Jones and aircrew) had one shot left and we had been on the ground for a few hours," Brown said. "I gave my own initials and cleared him hot."

Jones released the Hellfire missile and successfully destroyed the bunker, which allowed U.S. forces on the ground to recuperate and devise a mission plan going forward.

"When that Hellfire went into that bunker, beyond a shadow of a doubt, that bunker had been neutralized," Brown said.

The enemy may not have seen the MQ-1 as it soared overhead, but radical terrorists felt the Predator's wrath.

Jones and the rest of the MQ-1 aircrew loitered above the combat zone for approximately 14 hours, relaying critical information and laser-guided munitions during the entire fight. Their actions provided key reconnaissance for senior leaders commanding the situation, and directly enabled visual relay between forces on the ground and the combatant commander.

In the end, Gabe Brown defined the role of the RPA in this instance, and—in a broader sense—in the Global War on Terror from that point on: "I credit that pilot, the technology and that airframe with saving my life, as well as the team's, and getting the wounded and KIA (killed in action) off the hilltop that day."

Two service members received a Medal of Honor for heroic actions on Roberts Ridge, U.S. Air Force Combat Controller Technical Sergeant John Chapman and U.S. Navy SEAL Master Chief Petty Officer Britt Slabinski. Chapman's award was upgraded from the Air Force Cross to the Medal of Honor posthumously on August 22, 2018, in large part, according to official Air Force statements, due to the Predator's video that captured his heroism on that day.

Chapman, initially thought to have been killed in the first four minutes of fighting on the ridge, reemerged seventy minutes later on a Predator video and single-handedly engaged enemy positions on the ridge. Chapman was mortally wounded as a quick reaction force was inserted by a helicopter on the ridge. However, his actions were attributed with saving the lives of the reaction force. Technical Sergeant John Chapman became the first airman to receive the Medal of Honor since the Vietnam War and only the nineteenth airman ever to receive the medal. It is significant that video feed from an Air Force Predator served as a witness to Chapman's heroism and enabled the United States to award the highest honor to this warrior for his ultimate sacrifice.

Rise of the Robots

It should be noted that despite the current unappeasable demand for RPAs, in the early 2000s the military did not want to be involved in RPAs on a large scale. At that time, the military viewed RPAs as an unreliable platform with a niche mission set and an unproven track record. The Department of Defense's Unmanned Aerial Vehicle (or UAV, as the RPA was called at the time) Roadmap released in April 2001 illustrated the reluctance to adopt RPAs. Military priorities are evidenced by planned budgets to spend future money. Road maps are tools to inform the budget-cycle process and are not prescriptive in nature, but rather more descriptive of what is possible capability-wise.

According to the DoD's UAV Roadmap in 2001, $3 billion was spent on research, development, and procurement of RPAs from 1990 to 1999. The Roadmap projected that another $4 billion would be spent on RPAs by 2010 and that the number of RPAs in the overall inventory would rise from ninety in 2001 to 290 by 2010, a 322 percent increase in assets. At the onset of Operation Enduring Freedom in 2001 and the Battle of Roberts Ridge in March 2002, the United States Air Force had only two RQ-1 Predator squadrons and four RQ-4 Global Hawk aircraft. The United States Navy had only three RQ-2 Pioneer systems with five aircraft each. The United States Marine Corps had only two RQ-2 Pioneer systems with five aircraft each. And the United States Army had only four low-rate initial production RQ-5 Hunter systems with eight aircraft each. The DoD UAV Roadmap in 2001 further specifies the insignificant impact UAVs had on the military in April 2001:

Today's UAVs compose 0.6 percent of our military aircraft fleet, i.e., there are 175 manned aircraft for every unmanned one in the inventory. For every hour flown by military UAVs, manned military aircraft fly 300 hours. UAVs currently suffer mishaps at 10 to 100 times the rate incurred by their manned counterparts. UAVs are predominantly relegated to one mission: reconnaissance. Before the acceptance and use of UAVs can be expected to expand, advances must occur in three general areas: reliability, survivability, and autonomy. All of these attributes hinge on technology.

Even the MQ-1 Predator (in retrospect considered one of the most successful RPAs ever developed and operated) struggled to find ownership and advocacy in its nascent stages. The Predator program was originally managed by a U.S. Navy Program Office and was initially flown over Bosnia by a U.S. Army Military Intelligence Battalion (Mibli). Richard Whittle, in his book *Predator: The Secret Origins of the RPA Revolution*, described how the Predator came to be a USAF program:

...in April 1996, Defense Secretary William Perry agreed to give the Predator to the Air Force, but with a caveat. In a nod to "jointness," the ever popular idea that the armed services should cooperate and share hardware, Perry decreed that while the Air Force would operate the Predator, the Navy would keep developing the drone and procuring its various elements. And as a practical matter, the Army's Mibli would continue operating the Predator in Bosnia for a few more months while the Air Force trained some of its own pilots to fly an aircraft by remote control.

By 2010, the budget landscape within the DoD definitely changed in favor of unmanned systems. An updated version of the DoD UAV Roadmap from 2001, written in 2010 and referred to as the DoD Unmanned Systems Integrated Roadmap FY2011-2036 (referring to future fiscal years of the military budget) reflected that the services planned to spend $30 billion on RPAs between 2011 and 2015 and another $1.8 billion on ground and maritime unmanned systems. Bear in mind that what appeared to be a plan for lavish spending on unmanned systems in 2010 actually came after a recession in the United States and was during a time of fiscal constraint within the DoD budget.

The 2011 road map also described the unanticipated swelling of the DoD's unmanned aircraft inventory, which consisted of 107 Reapers and Global Hawks, 220 Predators, Gray Eagles, Hummingbirds, and Fire Scouts, 413 Shadows and Hunters, 122 Scan Eagles, and 6,810 small hand-launched RPAs, for a total of 7,672 total unmanned aircraft. The projected 322 percent increase in systems by 2010 actually materialized as an 8,500 percent increase (twenty-six times greater than anticipated)! The measly flight hours flown in 2001 increased significantly by 2010, logging 500,000 combat hours in 2009 alone, solely in support of operations in Iraq and Afghanistan. By November 2010, that number grew to one million total combat hours logged. Further evidence the RPA landscape changed appeared in the 2011 Unmanned Systems Integrated Roadmap, and extolled RPAs' importance:

U.S. and allied combat operations continue to highlight the value of unmanned systems in the modern combat environment. Combatant

Commanders (CCDRs) and warfighters value the inherent features of unmanned systems, especially their persistence, versatility, and reduced risk to human life. The U.S. military Services are fielding these systems in rapidly increasing numbers across all domains: air, ground, and maritime. Unmanned systems provide diverse capabilities to the joint commander to conduct operations across the range of military operations: environmental sensing and battlespace awareness; chemical, biological, radiological, and nuclear (CBRN) detection; counter-improvised explosive device (C-IED) capabilities; port security; precision targeting; and precision strike. Furthermore, the capabilities provided by these unmanned systems continue to expand.

As further evidence of the increased demand and usage of RPAs, during George Bush's eight-year tenure as president (2001–09) there were fifty-seven RPA strikes conducted. During Barack Obama's eight-year tenure (2009–17) there were 563 RPA strikes, a tenfold increase. In 2018, the Air Force manned more than sixty-five continuous flights (known as combat air patrols, or CAPs) worldwide, twenty-four hours a day, seven days a week. The number of CAPs would have risen to seventy but in 2015 Secretary of Defense Ashton Carter recognized the system was stressed to its limit due to manning issues and prevented the expansion.

When the DoD released the 2018 version of the Unmanned Systems Integrated Roadmap, it stopped bragging about numbers of RPA systems fielded and the millions of hours flown. It no longer needed to. RPAs were fully integrated into DoD operations. Instead, the 2018 road map simply stated, "To ensure our military advantage, emphasis should be placed on the evolution, availability, and employment of unmanned technology."

Although no country kept pace with the U.S., the RPA expansion was global. W. J. Hennigan described in a *Los Angeles Times* article, "A Fast Growing Club: Countries That Use RPAs for Killing by Remote Control" that "a total of 78 countries now deploy surveillance drones" and more than twenty countries "either have or are developing armed drones."

Commenting on this global expansion of armed RPAs, Peter Singer, author of *Wired for War*, stated, "It is a good illustration of how this technology has gone global. What was recently considered abnormal is the new normal of technology and war." Historical trends indicate that RPA usage will continue to globally increase in the future.

When Is the Appetite Sated?

An enormous demand for RPAs resulted from the U.S. fighting simultaneous wars in Afghanistan and Iraq. It would seem logical that as the wars wound down, the demand for RPAs would also decrease. But this hasn't been the reality. Instead, geographic combat commanders (the four-star general or admiral in charge of U.S. military operations in a geographic location such as the Middle East) requested more RPA support as the security situation eroded within their area of responsibility after the end of major hostilities in Iraq and Afghanistan. The Arab Spring in 2011 and the rise and rapid global expansion of the Islamic State in Iraq and Syria (ISIS) in 2014 further exacerbated existing demands for RPAs that could provide intelligence and persistent strike capabilities. Consequently, RPA operations expanded to more locations, such as Africa, Syria, and the Philippines, while remaining firmly planted in new and active campaigns in Iraq and Afghanistan.

New requests for RPA support were in addition to previously existing requirements, and there was no respite for organic RPA tasking despite the reduction of deployed forces on the ground in conflict areas. The military surged to support this "new normal." In *Kill Chain: RPAs and the Rise of High-Tech Assassins*, Andrew Cockburn described this exigent appetite well with a quote from General James R. Clapper, a prominent figure in the defense intelligence community including time served as the director of national intelligence, director of the Defense Intelligence Agency, undersecretary of defense for intelligence, and director of the National Geospatial-Intelligence Agency. Clapper, referring to RPAs, said, "I cannot see a situation where someone is going to say, 'Hey, I can do with less of that.'" Which brings us to where we are today and will most likely remain in the future: a never-ending hunger for RPAs with no appetite suppressant in sight.

Manning the Unmanned

The insistent demand for RPA support is not, however, without repercussions. The rapid growth of RPAs, over a relatively short period of time for the military, created several unintended consequences, workarounds, shortcuts, and areas that were immature in thought, readiness, and process. As evidence of the quick expansion, in 2011, a decade after the catalyst that sparked the RPA demand, the Air Force trained one hundred more RPA pilots annually than traditional pilots. While a decade may seem like a long time to adjust to this newly emerged requirement, consider first that it was an unanticipated requirement born out of necessity at the crossroads of technological advancement, budget cuts, and increasing demand by combatant commanders and frontline troops. Also consider that prior to January 2011, RPA pilots in the Air Force were traditional aviators pulled (most times begrudgingly) from their previous aircraft into the RPA community to fill the demand.

In an article titled "A Chilling New Post-Traumatic Stress Disorder: Why Drone Pilots Are Quitting in Record Numbers," Pratap Chatterjee described the Air Force's solution to manpower shortages in the RPA community: "The Air Force has had to press regular cargo and jet pilots as well as reservists into becoming instant drone pilots in order to keep up with the Pentagon's enormous appetite for real-time video feeds from around the world."

Aaron Church, associate editor of *Air Force Magazine*, described this demand in a June 2011 article titled "RPA Ramp Up":

The demand for the intelligence surveillance-reconnaissance capability offered by the UAVs continued to mount, however, and the new target of 65 RPA orbits by 2013—a 1,200 percent growth in operations since the war in Afghanistan began—rendered the ad-hoc manning scheme untenable. To operate around the clock, each orbit requires 10 aircrews, including a pilot and sensor operator. That means USAF needs an estimated 1,350 RPA crews by 2013, not including reserves, according to AETC (Air Education Training Command) estimates.

It became apparent the requirement for RPAs would not subside and the existing manning model within the United States Air Force (USAF) was insufficient to satisfy the future demand. As a result, in January 2011 the Air Force Specialty Code (basically a job title and description) of "18X—RPA Pilot" was created. The ramp-up to meet the audacious manning goal, coupled with the subsequent RPA support to enduring operations, inevitably produced some unintended consequences that took years to surface.

One of the most prominent of these unforeseen issues identified was the psychological responses to killing experienced by the crews operating from half a world away. Understanding how and why RPA personnel respond to killing will be paramount not only for the RPA enterprise as a whole but for the health and well-being of the men and women operating around the clock with little reprieve in an attempt to satiate the ever-hungry construct that has become unmanned operations.

Chapter Three

The Machines

Drones are just another weapon, and they turn out to be a very effective weapon that puts no American troops at risk, and I don't see why we shouldn't use them against identified enemy targets.

—Colin Powell

RPA Familiarization

To better understand killing with RPAs we must first be familiar with the machines themselves: their capabilities, as well as how they are employed, categorized, and manned. In this chapter, I'll also include comments drawn from interviews conducted with the pilots who have flown each RPA discussed.

First, let's start with what RPAs are called. The term *drone* is a common moniker used to describe any remotely controlled unmanned aircraft regardless of size or capability. Within the military community of people trained to fly, operate, or maintain unmanned aircraft, drones are known by different, more technical names such as unmanned aircraft system (UAS), remotely piloted aircraft, remotely piloted vehicle (RPV), or in some cases, although the term's usage has lost favor, unmanned aerial vehicle.

I chose to use the acronym RPA throughout this book as a common point of reference to describe unmanned aircraft for two simple reasons: It is the technical term widely recognized in the military to describe a remotely controlled, unmanned aircraft, and it is brief, simple, and clear. The term *drone* is actually a pejorative description and has negative connotations for military personnel associated with unmanned aircraft for reasons I discuss in the chapter on culture. That is not my intent any time I deliberately choose to use the word *drone* and no slight is implied by its use in this book.

One downside to using a single term to encapsulate a wide variety of unmanned aircraft is that it lumps the vast spectrum into one category regardless of differences and blurs the lines between civilian and military use. The word *drone* invokes two thoughts: an annoying quadcopter flying overhead and invading someone's privacy, or an armed remotely controlled military aircraft used to strike a target with a missile. In reality, those are just two of many types of drone that exist in the vast inventory of RPAs.

The differences between RPAs, however, are extremely relevant in the military due to the wide range of capabilities and how the RPAs are used either to facilitate killing or to kill directly. To fathom the differences among RPAs we must first understand how they are categorized, flown, and used. In 2010, the Department of Defense categorized RPAs into five distinctive groups based on maximum gross takeoff weight, nominal operating altitude, and speed. Prior to 2010, each branch of the U.S. military described its unmanned aircraft according to a service-generated tier system.

Categories of RPAs

Group 1: Small UAS

Group 1 contains the smallest RPAs, weighing twenty or fewer pounds, flying no higher than 1,200 feet above the ground, at a speed of less than one hundred nautical miles per hour (knots). Group 1 RPAs encompass a wide variety of civilian hobby and commercial drones as well as what the military refers to as "smalls" or small UASs (sUAS). In general, small RPAs take off and land independent of a runway (in some instances they are thrown in the air by the pilot to take off), fly for approximately an hour, are flown by a single pilot within line-of-sight range of the airplane, and are limited in range to a few miles.

But even within the category of small RPAs, there are outliers that fly for three hours out to a range of almost forty kilometers. Small RPAs are powered by rechargeable batteries or fuel cells as opposed to the combustible engine common among larger RPA systems. Fuel cells offer the greatest endurance and range but are more expensive and less prominent in use. A global positioning system (GPS) flight plan can be programmed into the flight control logic of the sUAS to assist with operator workload, enabling the single operator to focus on manipulating the onboard camera instead of flying the RPA. Small RPAs carry an onboard camera capable of zooming in on objects of interest and are predominantly used for surveillance within a localized area by frontline troops. Very minimal training (usually less than a month) is needed to operate a small RPA.

Because of the ease of use, low cost, and ubiquitous nature, small commercial off-the-shelf RPAs have been militarized throughout the world by state and non-state actors. We have entered a gray zone in warfare where the technologies of war can no longer be controlled by the state but rather purchased at a local store or through the internet and then transformed for military purposes with relatively little effort. We see this playing out in conflicts throughout the globe as commercial drones increasingly appear on the battlefield.

Despite a limited capacity to lift additional payloads, small RPAs have been modified on occasion to carry very light munitions that are dropped like a small bomb, or the RPA itself is flown kamikaze-style into the target with the explosives on board. The kamikaze method was demonstrated in the attempted assassination of the Venezuelan president Nicolás Maduro on August 4, 2018, by two drones carrying two pounds of explosives each.

In 2016, the Islamic State modified commercial off-the-shelf drones to drop grenades and mortars on troops, equipment, and buildings in Iraq and Syria. Subsequently, the Islamic State claimed to be the first terrorist group to successfully weaponize a small drone and kill with it. In Ukraine, crowdfunding from the civilian populace aided the development of small commercial drones to support the Ukrainian Army's fight against Russian-backed separatists. The upsurge of cheap, commercial off-the-shelf drones meant that anyone could field an "air force" on the modern battlefield.

Marine Special Operations Command (MARSOC) led the way within the U.S. military using small lethal UAS. In 2010, MARSOC owned only five RQ-11 Raven systems specifically used for surveillance. Four years later, when the U.S. withdrew large numbers of troops from Afghanistan, Marine special operations teams were distributed throughout the globe to assist foreign militaries in the fight against their enemies.

What didn't follow the special operations teams in the global re-assignment was sufficient direct support RPAs to provide intelligence, surveillance, and reconnaissance (ISR) and precision strikes for deployed teams. Out of necessity, MARSOC built a family of small UASs and filled the void internally to operate in every environment. Part of that small UAS family of systems included what is called lethal miniature aerial munitions (LMAMs), with the most prevalent one named "Switchblade" from the manufacturer AeroVironment.

LMAMs, as the name implies, are a kamikaze-like UAS with a small warhead capable of loitering for a short duration (approximately twelve minutes) out to a range of ten kilometers. Switchblades are launched from a tube on the ground (similar to a mortar) and steered in flight to the target using an onboard camera that sends video back to the operator. Switchblades receive information, such as target location, directly from other small UASs within the family of systems and provide unprecedented organic precision strike capability down to the lowest tactical level on the battlefield. MARSOC's success employing Switchblades paved the way for the use and proliferation of small lethal UASs within the regular military, particularly the Army and Marine Corps.

The U.S. military's dominance employing LMAMs was short-lived. During the Nagorno-Karabakh conflict between the Armenians and the Azerbaijanis in 2020, the Azerbaijanis flew Israeli-made Orbiter 1K kamikaze drones with great success.

Three factors to consider concerning the psychology of a small lethal UAS operator that differentiate it from larger RPAs are the close proximity of the operator to the target, the limited persistence of the RPA, and the primary function of the operator employing it. Due to the limited range, operators of small lethal RPAs must be forward deployed, less than ten kilometers away from the target. The close proximity to the target implies a certain level of risk or danger to the operator, which doesn't allow for a like comparison to larger RPAs. Due to small lethal RPAs' limited endurance, missions such as observing a target for a long period of time prior to and after striking it are not accomplished with the RPA itself, so no intimacy with the target is developed by the operator. Finally, the operators of small lethal RPAs are either infantrymen or special operations troops first and are incidentally trained to operate RPAs second.

For an infantryman or special operations critical skills operator, the farther out one can kill a target, the less personal risk and exposure to danger the operator experiences. It therefore follows that employing a lethal RPA from a range that keeps the operator farther from harm's way is psychologically preferable to having the warrior directly assault the target with traditional infantry weapons. As such, using small RPAs to kill actually produces positive psychological responses for operators because it keeps them out of more "kill or be killed" situations. Consequently, despite the fact that operators of small lethal RPAs are using these weapons to kill from a relatively short remote distance, they have reported no negative psychological responses to killing with small RPAs. One member of MARSOC, who chose to remain anonymous, stated in an interview:

> The natural question with these small things that fly is how can we use them to kill the enemy and break his stuff. When you look at the guys flying these small lethal RPAs it is Marines, Army, and special operations;

they are trained to kill first before they are trained to fly an RPA. They carry a rifle for a living. It's relatively close range, quick reaction, real-time killing. It's SOF [special operations forces] and infantry. It's a less dangerous form of killing than having to kick the door in.

Group 2: Tactical UAS

Group 2 RPAs weigh between twenty-one and fifty-five pounds, fly no higher than 3,500 feet above the ground, at a speed of less than 250 nautical miles per hour (knots). Group 2 UAS are predominantly catapulted from a pneumatic launcher to take off and remain aloft upward of twenty-four hours out to a range of one hundred miles, depending on the model. They are capable of flying global positioning system (GPS) flight plans, carrying limited payloads in addition to the onboard camera, and are used for ISR. The historically predominant Group 2 systems are Scan Eagle and Aerosonde, routinely employed as a contractor-owned, contractor-operated, fee-for-service system, meaning the military "leases" camera hours from the equipment manufacturer, who flies and maintains the RPA in support of forces on the ground or on a ship.

The genesis of the leasing of camera hours, as previously discussed, was the lack of organic RPAs within the U.S. military to fulfill the demand during wartime. The Group 2 platforms that produced the leased camera hours were flown by civilian contractors and presented legal challenges, since international law prevents civilians from directly participating in combat operations. In order to comply with international law, the military drew the line at flying armed missions with civilians at the controls. Therefore, no Group 2 UASs were armed. But that constraint did not prevent the use of these aircraft to search for legitimate military targets. To aid in that searching process, military liaisons, such as military intelligence personnel and RPA pilots, were assigned to contractor crews for each mission to convert "leased" camera hours into actionable information.

Marine Captain Douglas Wood was one of the intelligence officers deployed to Iraq in December 2016 as part of Task Force Al Assad. This task force provided some of the critical assets the Iraqi Army needed to defeat the Islamic State, including ISR and precision fires. As part of the task force, Captain Wood had three RPA systems assigned to him, including a Group 2 Scan Eagle system that was used to hunt for Islamic State fighters. Once a legitimate target was identified by Captain Wood or his Marines, it was nominated as part of a target package and sent to the Target Validation Authority (TVA) in Baghdad for approval to strike. After the target was positively identified by the Scan Eagle and approved by the TVA, coalition aircraft were then called in to strike the target.

The Iraqi base of Al Assad, once a sprawling metropolis of coalition military forces during Operation Iraqi Freedom, was completely surrounded by the Islamic State in 2016. As such, there was never a lack of enemy fighters to target. During Captain Wood's nine-month deployment to Iraq, he and his team facilitated more than 650 strikes using RPAs as the spotter, resulting in more than eight hundred Islamic State members killed in action. Two hundred fifty strikes were deliberate targets, such as buildings housing enemy fighters; the remainder were dynamic (situational) strikes, such as identifying enemy fighters in the field. Although he was remote from the RPAs on the battlefield that piped video back to him at Al Assad, Captain Wood described in vivid detail how he experienced the humanity of the targets he struck:

> You are getting this really intimate look into a person's life because the cameras are so good. We watched a guy emplace IEDs for thirty minutes and then go home and have a fight with his wife as he constructed more IEDs in his backyard. It was a very humanizing experience. It's incredibly voyeuristic.

Despite recognizing the humanity of his targets, Captain Wood felt no compunction for killing Islamic State fighters at the time. This was in large part because, as an intelligence officer, he digested myriad information about the horrific acts the Islamic State committed, such as burning and beheading civilians. But the killings weren't without a cost on his psyche even though he wasn't at the controls of the RPAs and directed their employment from the safety of the base. Over the course of the nine-month deployment, Captain Wood remained at a heightened level of stress due to the numerous killings he observed. His daily hypervigilance manifested over time and he became angrier and more physically aggressive, and experienced trouble sleeping.

His team shared the same experiences and responses. While Captain Wood was not exposed to physical danger, excepting the occasional indiscriminate missile attack on the base, his body responded to extreme stress as if he were in danger. This stress was evident even to his wife upon his return from deployment. But, once removed from the situation, the stress subsided over time. Two years removed from the situation, Captain Wood still lucidly recalled some of the horrific events he observed from a Group 2 UAS video, such as the time he watched six Islamic State fighters burn alive when a fuel tank exploded after being ignited by a Hellfire missile. Captain Wood, an extremely introspective person, displayed raw emotion during my interview with him, and it was apparent that the combat he participated in and witnessed lingered vividly in his mind.

Group 3: Tactical UAS

Group 3 RPAs weigh less than 1,320 pounds and fly below eighteen thousand feet at a speed of less than 250 knots. They share characteristics similar to Group 2 UASs but have an increased capacity to carry payloads in addition to their onboard cameras. The historically predominant platforms within the Group 3 category have been the RQ-2 Pioneer, RQ-7 Shadow, and the RQ-21 Blackjack flown by the Navy, Army, and the Marine Corps. The R in the naming nomenclature stands for reconnaissance, while the Q stands for unmanned aircraft. Group 3s are considered a tactical-level asset, meaning they are employed at the tactical level directly in support of ground troops or service-specific tasks and don't receive mission tasking from the commander in charge of all the aviation assets during an operation (known as the Combined Forces Air Component Commander [CFACC—pronounced "See-Fack"]). Employment at the tactical level also implies that Group 3s are flown in close proximity (less than one hundred miles) to their targets and the supported friendly troops are within radio line of sight of the crew on the ground.

The crew consists of a pilot to fly the plane and a sensor operator to manipulate the camera. Depending on the military service and situation, the crew could also include a mission commander and intelligence analysts. Group 3s are primarily reconnaissance platforms and due to physical flight performance limitations don't carry any onboard munitions. They are capable of carrying onboard radios to communicate directly to ground forces and laser designators to guide laser-guided munitions onto a target. Although Group 3s are not armed, they do frequently aid other assets (such as artillery and attack aircraft) to strike targets.

Marine Staff Sergeant Christopher Herr flew Group 3s for thirteen years; RQ-7 Shadows in Afghanistan and RQ-2 Pioneers in Iraq before that. Staff Sergeant Herr recounted several instances of flying in 2010 in Afghanistan, when he deployed with Marine Unmanned Aerial Vehicle Squadron Three (VMU-3) and participated in Operation Moshtarak, the battle for the town of Marjah, the last stronghold of the Taliban within Helmand Province. For two weeks prior to coalition forces seizing the city, Herr and his fellow Marines flew Shadows over the city, day and night, at a low altitude where the RPA's engine made a very distinctive, loud noise similar to a lawnmower that could easily be heard on the ground below. However, the Marines used the Shadow's noise to their advantage, as a psychological weapon, to disrupt the enemy's sleep and desensitize them to the sound of RPAs that were later used for target identification.

Once the battle commenced, during one flight Herr identified a man in his camera splashing water around some holes in the exterior wall surrounding a house. What appeared to be an innocuous act, Herr rightfully interpreted as an enemy fighter preparing a firing position, attempting to reduce the amount of dust signature that would appear when he fired his rifle at U.S. troops through a "murder hole" (a small hole in a wall used to shoot at the enemy and conceal your position). Once the man displayed a rifle, Herr marked the wall next to the enemy fighter with a laser from his RPA and called in an AH-1W SuperCobra attack helicopter to fire a Hellfire at the laser spot, killing the man. While Herr admitted that was probably the best day of his deployment, he also recounted one of his worst days:

> I once spent five hours flying at two thousand AGL [above the ground level] in Afghanistan, under a cloud deck on a rainy day, just to surveil the corpse of an enemy killed earlier that day to ensure nobody would later come to retrieve the body and if they did to allow us to follow them to their hideout. We were low enough due to the cloud cover that everyone within three kilometers would have been able to hear the noise from our engine and thus stayed indoors, but common sense was not a common virtue and we were forced to sit there watching the slowly fading infrared signature of a dead body for no reason.

Observing a dead body for five straight hours is an abnormal event that very few people other than morticians have experienced, unless you work with RPAs. Herr admittedly didn't know if he would experience any long-term issues from his experiences flying RPAs in combat. He, like so many involved with RPAs, chose to leave the Marines because of his perception of a negative culture toward RPA pilots.

Group 4: Medium Altitude Long Endurance

The next largest group of RPAs look more like traditional airplanes and helicopters. Group 4 RPAs, often referred to as medium altitude long endurance (MALE), weigh more than 1,320 pounds and fly up to an altitude of eighteen thousand feet. Historically predominant platforms in this group are the Navy's MQ-8 Fire Scout (a helicopter), the Air Force's MQ-1 Predator, and the Army's MQ-1C Gray Eagle (which is a Predator built for the Army but capable of carrying four Hellfire missiles instead of two). The M stands for multi-role. Upgrades from Group 3s include improved aircraft reliability, increased range, capacity to carry larger payloads and munitions (particularly Hellfire missiles), and the ability to fly via satellite communications. Group 4 RPAs are flown by a single pilot, with a sensor operator controlling the cameras, but require upward of two hundred people per squadron to maintain, fly, and analyze the information collected. Primary missions for Group 4 RPAs are reconnaissance and strike, with endurance lasting from ten to twenty-four hours, depending on the RPA.

The larger, armed RPA is significant for two major reasons: the maximum distance from which the crew can fly it over a satellite link and the inclusion of ordnance on the plane, meaning the crew is killing directly from its own aircraft. An enlisted Army MQ-1C Gray Eagle pilot with the 160th Special Operations Aviation Regiment who chose to remain anonymous recalled his first shot while deployed to Afghanistan. It's important to note that, while technically possible, the U.S. Army traditionally chooses not to fly its RPAs via satellite from the U.S. Instead, it forward deploys the crews to the area of conflict.

The unit before us only shot three times during their deployment. We were in the middle of our RIP [relief in place] with that unit and I was the first person set to be qualified as an aircraft commander within the unit. I only needed one more qualifying event to be an aircraft commander, which was to shoot a live Hellfire missile at a boulder outside of our FOB [forward operating base]. Once I fired that training shot and completed the qualification, I was signed off to fly live missions by myself and continued on with the flight that day.

Shortly after shooting that rock, I was dynamically re-tasked over chat to go and support a TIC [troops in contact]. The TIC was a MARSOC [Marine Special Operations Command] unit in a pretty deep valley that was pinned down. On the high ground terrain above the MARSOC team were two sets of three insurgents firing down on them with rifles and RPGs [rocket-propelled grenades], and the situation went from me shooting a live missile for the first time an hour earlier to getting a nine-line [a request to strike a target from a ground force] to support this team being shot at. I was super green at that time and freaking out a bit.

The two groups of three insurgents were too far spread out for us to take a shot on both of them, so I lined up for a shot on one group. It was very stressful because the Marine JTAC [Joint Terminal Attack

Controller] understandably was screaming on the radio for help. Once we found them [the enemy] we turned in for the shot and I made a mistake and the sensor went nadir [too close overhead of the target to remain locked on to it with the sensor] and we had to turn around, which raised the stress level of the JTAC, who began screaming over the radio, "We need you right fucking now!"

At that same time, my mission commander stopped into the GCS [ground control station, the RPA's cockpit on the ground] and told us we weren't approved to take the shot. I told him we were going to shoot because the JTAC needed it. When the mission commander left to go get his commander, I got up and locked him out of the GCS so I could continue with the mission.

As my stress level rose, I reset the airplane and took the shot. We saw a plume of smoke, then an explosion, and then three bodies lying on the ground. That shot scared off the other insurgents and they stopped shooting at the MARSOC team, who was able to withdraw with minimal casualties. What made that first shot more impactful for me was that MARSOC team, which was a month out from leaving the country at the time of the incident, later came to our base to see us.

The JTAC (whom I talked with on the radio) pulled me into a room and broke down crying as he stated, "That was my worst day in theater. I have a wife and kids and I didn't think I was going home. That was the first kinetic strike I got to do, and I was scared that we weren't going to make it out of that situation until you showed up."

I ended up having thirteen shots during that deployment and each time I had the same physical responses. A minute of time felt like twenty minutes. My hands got cold and felt like they were dipped in buckets of ice water. After the shot, I felt my hands slowly start to get warmer as the blood rushed back into them. I would also notice that sometimes I wouldn't be breathing because I was so focused on the shot and I had to do some meditative breathing to regain control of my breathing.

After each shot, I got this strange sense of elation, almost comparable to that feeling after sex. I never had any negative responses to strikes, though. It makes me feel weird to say it, but those kinetic strikes were some of the best experiences of my life. Not a one that I had was ever in question if I was shooting a good guy or bad guy. If they deserved it or not. Each and every shot I had was either troops in contact or someone we were watching actively preparing to attack our guys.

Several things are at play in this story that I will discuss in detail in other chapters: training during combat, connection to friendly forces, justification and rationalization of killing, and psychological and physiological responses.

Group 5: High Altitude Long Endurance

Group 5 RPAs are the largest and most capable RPAs, often referred to as High Altitude Long Endurance (HALE) platforms. They weigh more than 1,320 pounds and aren't defined by a particular airspeed or altitude limitation. Group 5 includes the most sophisticated and expensive platforms developed, such as the Air Force's RQ-4 Global Hawk and MQ-9 Reaper and the Navy's MQ-4 Triton. Global Hawk and Triton are used as operational and strategic-level surveillance aircraft. They carry a wide array of sensors, including daytime and infrared cameras as well as payloads that intercept radio signals, and synthetic aperture radars (which permit the sensor to develop an image of objects below based on radar returns regardless of bad weather or cloud cover). Global Hawk and Triton can fly for more than forty hours out to ranges greater than fourteen thousand miles. Although impressive in capability and extremely useful for strategic-level intelligence collection and target development, Global Hawk and Triton do not participate in real-time strikes at the tactical level and their crews are therefore outside the scope of my study.

The Predator's replacement, the Reaper, on the other hand, is *the* most prevalent armed RPA in the U.S. inventory. The Reaper can carry fifteen times more payload than the Predator, while flying three times faster with its turboprop engine. Reapers carry a plethora of bombs and missiles, including laser-guided Hellfire missiles, GPS- or laser-guided bombs, a high-definition day/night camera, as well as mission-specific special payloads. Reapers can fly a maximum speed of 260 knots, have an endurance of fourteen hours, with a range greater than a thousand miles, up to an altitude of fifty thousand feet. Similar to the Predator, the Reaper can be controlled over a satellite radio communication link and requires a large support structure of personnel outside of the crew.

An Air Force Reaper pilot described to us what it was like to strike a target in Afghanistan remotely from Nevada over that satellite link:

> I had one shot where we hit the guy. There's this explosion of dirt and dust and the camera gets fuzzy for a minute and it's hard to see what it all looks like. We actually thought we missed because when the smoke cleared all there was left was this smoking hole. We thought he must have gotten away or rolled away down into a ditch. So, we started looking for him.
>
> We never found him until some villagers came by. They came in flocks; ten people at first, then thirty, then sixty, then one hundred. They brought sheets with them and they started throwing little things onto the sheets that looked like they were picking up rocks. Then they started peeling things off of a wall and throwing it onto the sheet. And then we realized it was body parts of this guy that had just been completely obliterated. Watching them do that, and coming to the collective realization of what it was, quieted the room. We watched that scene for two hours. And then followed the villagers back to the funeral preparations area and watched that for another hour or two. Earlier in the day we actually found that high-value target at the beginning of my flight when he was attending a funeral for someone else and then wandered off by himself afterward where we struck him outside that village.

Upon completing his flight that day, the Reaper pilot got in his car and drove home to sleep in his own bed, mentally and physically exhausted from the day's events and the additional administrative paperwork post-shift that results from a strike. He would awake the next day and commute to combat again, where, for five days in a row, he killed someone on the other side of the globe. It is the humans who fly these aircraft who are the most important part of this system, as I discuss in the next chapter.

Chapter Four

The People

War is a human endeavor—a fundamentally human clash of wills often fought among populations. It is not a mechanical process that can be controlled precisely, or even mostly, by machines, statistics, or laws that cover operations in carefully controlled and predictable environments. Fundamentally, all war is about changing human behavior. It is both a contest of wills and a contest of intellect between two or more sides in a conflict, with each trying to alter the behavior of the other side.

—Army Doctrine publication 3-0, Operations

We Weren't Ready, but "Off We Go into the Wild Blue Yonder"

War is an ugly, impatient beast that persists regardless of man's preparedness for it. To say that the U.S. military was unprepared for exponential expansion of the RPA community at the onset of war in 2001 would be an understatement. Then nearly a decade passed before the process became formalized. In 2011, the U.S. Department of Defense was still plugging gaps in RPA personnel requirements from within the institution, in most cases begrudgingly pulling pilots out of the cockpit of their traditional aircraft and forcing them to fly RPAs. Internally growing a force of trained, professional RPA pilots and sensor operators became fait accompli when the global demand for RPAs showed no sign of relenting. Forced into constraint with both limited tools and space to burgeon, each service went about growing its own garden in a different manner.

The Air Force had the most extensive and comprehensive plan to grow and train its RPA force. The plan started with the creation of occupational fields for sensor operators in 2009 and RPA pilots in 2011. To ensure these newly designated personnel had a training curriculum, a training pipeline specific for RPA personnel that resembled training for manned aviation was implemented. The "pipeline," as the RPA training program is referred to in the Air Force, consists of forty hours of initial flight training in a single-propeller airplane (along with their manned counterparts), instrument qualification in a T-6 simulator, an RPA tactical fundamentals course, as well as formal training in the specific aircraft assigned, such as the MQ-9 Reaper.

While modeled after a traditional aviation training regime, the pipeline differs in one major way. Since Air Force RPA squadrons only support live combat missions 24/7, RPA personnel serve as the pilot in command or qualified sensor operator and contribute to the fight in their first operational squadron on day one of their arrival, unlike their manned counterparts, who train up to the point of being wingmen or copilots when they arrive at their first operational squadron. When newly minted RPA pilots and sensor operators arrive in their line unit, any remaining mission training occurs during live combat missions. Institution of this trial-by-fire methodology is neither intentional nor desired but is the status quo largely as a result of insufficient time and resources to train due to the demanding operational tempo.

The Air Force also leads the DoD in the screening process for RPA personnel accessions. In the article "Aptitude and Trait Predictors of Manned and Unmanned Aircraft Pilot Job Performance," Laura Barron, Thomas Carretta, and Mark Rose describe the personality traits and knowledge that predict success as a traditional or RPA pilot. For the study, the team reviewed first-year performance evaluations of 3,140 Air Force traditional pilots and 330 RPA pilots. In addition, performance evaluations were reviewed over a period of three years for 1,662 manned pilots and 146 RPA pilots. The study revealed that the qualifications and personality traits that predict successful job performance for a manned or remote pilot are near identical, with one exception.

Greater pre-accession general aviation knowledge (meteorology, aero-dynamics, navigation, aircraft parts, and terminology) for RPA pilots is a high indicator of their future success. The authors suggest preexisting aviation knowledge, summed up conceptually as passion, is an indicator of success and pilots that lack this passion comparatively experience degraded performance as the stresses of being an RPA pilot emerge. In other words, if your heart's not in it from the beginning, being an RPA pilot doesn't make it any better and your performance declines as stress factors increase, such as long shift work or the ever-present demand to strike targets.

Although screening and recruitment of USAF RPA personnel has not differed from its manned counterparts, one study that researched the psychology of weaponized RPAs suggests that the skills required to be a successful RPA pilot, such as processing multiple auditory and visual inputs, might actually require a distinct recruitment and screening process. So even though the Air Force is leading the way within the DoD in the recruitment and screening process, much work remains to be done to tailor this process to find the right warriors to operate these platforms.

Let's Talk About Killing

In the early stage of the Air Force RPA training pipeline, discussions are held on the likelihood that an RPA aircrew will be required to kill if assigned to the MQ-9 Reaper. I experienced this firsthand while immersed in the training pipeline and observed how one young Air Force lieutenant who expressed concern over whether he could take a life was chosen for the unarmed Global Hawk platform instead of the armed Reapers. While it's reassuring that the discussions are being held at this early stage, it's extremely difficult to predict successfully how people respond to killing. No one can know for certain, but the Air Force is getting it right by having these early discussions. As addressed in *On Combat*:

> The time to decide if you can kill another human being is not in the middle of combat. The time to decide, to the utmost of your ability, is right now. We must be able to know if someone is willing, capable, and able of killing someone. It should be part of the "interview" for the job.

At this point of inflection in the job interview, some Air Force students within the training pipeline actually choose to be reassigned elsewhere within the Air Force rather than face the possibility of having to kill as part of their occupation. That's not a hit on their character, it's a sign that the screening process is working.

Unmanned Hooah!

The U.S. Army trains its RPA personnel at Fort Huachuca, Arizona. The training differs from the Air Force model in two major ways. Army RPA pilots and sensor operator positions are both filled by enlisted soldiers, and there is no initial flight training in a traditional airplane. Otherwise, the curriculum and training remain similar to the Air Force and is conducted over a twenty-three-week period. However, unlike the Air Force, no discussions about killing occur with students during Army entry-level RPA training. During the Army training process, selection of either a Group 3 (unarmed) or Group 4 (armed) RPA is determined for students based on aptitude and performance, favoring the larger, armed RPA for students who perform well. Since both the pilot and sensor operator positions are filled by enlisted soldiers, the Army relies on training qualifications to determine which position a soldier performs during a mission. Upon completion of initial training at school, soldiers are trained on mission skills in their first unit, since they aren't required to fly a combat mission immediately upon arrival as aircrew in the Air Force are.

Unmanned Oorah!

The U.S. Marine Corps sends its officers through a portion of the Air Force training pipeline, up to the point of aircraft-specific training. The officers then attend service-led, aircraft-specific training in Cherry Point, North Carolina. For two decades, Marine Corps–enlisted operators attended the Army's training at Fort Huachuca, but that changed when the Marine Corps bought a service-specific RPA not flown by the Army. Historically in the Marine Corps, enlisted RPA operators both flew the plane and served as the sensor operator. That changed in 2012 when, similar to the Air Force, the Marine Corps created an officer RPA pilot occupation field and expanded in 2020 to include Marine MQ-9 Reaper pilots.

Prior to 2012, officers were pulled from within the aviation community (pilots, naval flight officers, and command and control officers), similar to the Air Force, and served in RPA squadrons as mission commanders, conducting external mission coordination and overseeing missions flown by the enlisted operators. Marine officers also transitioned from previous occupation fields into the RPA community through a transition conversion board process where candidates are selected by a board of people who review their performance record coupled with their background. Most selectees are prior traditional pilots or someone with a background in fires such as artillery or infantry. Pre-accession screening consists of achieving a suitable score on an aviation skills test battery, the same test given to all naval and Marine pilots and naval flight officers, as well as passing a flight physical.

No discussion occurs in the screening or training process about the inevitabilities of the job, particularly killing but, as discussed in the introduction, killing discussions permeate the entire culture in the Marine Corps. Every Marine, officer or enlisted, is trained to serve as a rifleman before receiving specific occupational training, and it's assumed in most Marine occupations that killing may be part of the job.

Unmanned Hooyah!

In 2020, the U.S. Navy created an occupational field for warrant officers to fly the MQ-25. The other Navy RPAs are flown by traditional manned aviators, usually helicopter pilots or P-3/P-8 pilots, who are trained to fly an RPA in addition to their primary aircraft. The Navy does not fly armed RPAs as of 2021 and traditionally doesn't employ its RPAs in a fire support capacity.

Who's Flying This Thing?

Now that we have gained an appreciation for the training and screening process of each branch of the U.S. military, let's focus on the humans who fly RPAs. War, after all, is a human endeavor and using RPAs to fight does not alter that fact. While RPA warriors may be remote from physical danger, they are not insulated from all that occurs in combat. In order to understand how and why people respond to killing with RPAs, we must first know the aircrew's roles, responsibilities, and method of employment.

One quick statement about other warriors before I move on to talk about RPA warriors. The discussion in this book about remote killing with RPAs is not intended to diminish the sacrifices and hardships endured by any other warriors, particularly those on the front lines risking their own personal safety and security to accomplish the mission. This isn't an either/or situation. It isn't a suffering competition. Everyone in the military has their role based on training, proximity to the fight, and occupational specialty.

The entire point of this book is to describe the unique situation created for our service members who fly RPAs and kill people from a remote distance. RPAs were the next natural progression in the development of tools of warfare that further distanced service members from the dangers on the battlefield and simultaneously still had a significant impact on a conflict. However, as we'll see, it isn't the physical distance from the fight that is most important when understanding what RPA warriors experience, it is their cognitive or empathetic distance from the fight.

Pilots

Let's start with the person in charge: the pilot. RPA pilots can be either enlisted personnel or officers, depending on the branch of service and the platform flown, whereas traditional manned aircraft pilots are either commissioned or warrant officers. Every RPA, regardless of size, has a pilot, and the training required for the pilot differs based on the RPA.

A small RPA doesn't require the same level of training to fly as a Reaper, but it still has a pilot in the sense that someone manipulates the flight controls of the aircraft. The pilot is normally in command of the flight, meaning they serve as the final arbiter for any decision that is made regarding how the mission is flown, what occurs during the mission, maneuvering the airplane, safety-related issues, and emergencies. The pilot in command is the ultimate, absolute authority over everything during the flight, the same as in a traditional, manned aircraft.

There are many common misconceptions about RPA pilots perpetuated in the media, the entertainment industry, and even within the military: that they are disengaged emotionally from their mission, that flying an RPA is like playing a video game, and that RPA pilots are all mentally and morally traumatized from the things they have done, to name a few. While there might be a small percentage of RPA pilots who are accurately described by these statements, we cannot construe these descriptions as accurately portraying the entire population.

On the contrary, RPA pilots are well-trained, professional warriors who strive to do their best, honestly care about the missions they have been given and the troops they support, and are physically and emotionally invested in achieving success. They *are* connected to the fight, regardless of physical distance from it. This assertion may seem counterintuitive, but several studies conducted by Air Force psychologists and my research confirm it. The absence of danger to oneself shouldn't be the basis for judging an entire occupation. The majority of military personnel in support roles and almost all civilian occupations likewise don't encounter dangerous situations, but no one makes disparaging comments about doctors or air traffic controllers for not risking their life for their job despite the life-and-death decisions they make on behalf of others on a daily basis. And, certainly, most people would not diminish the contribution of the many members of the U.S. armed forces whose service never sent them into the combat zone.

It's ironic that the very same population that conflates what is an honorable military profession with "playing a video game" is dominated by people who never face danger or are required to kill in their own occupation. Instead, we should celebrate the fact that due to the contributions of the RPA community, fewer of our sons and daughters face the possibility of suffering physical wounds or dying in battle. After all, isn't that part of the purpose of war—to kill the enemy, break his stuff, and bend him to our will while simultaneously trying to protect and preserve the infinitely precious lives of our own service members? To make the decision to go to war, and then not do everything in our power to ensure victory, and to protect the lives of our service members and the lives of innocents, can be seen as a tragic foolishness, and even an inexcusable dereliction of duty.

In recognizing the contribution of our RPA crew members, we do not diminish the roles of those who go in harm's way to serve our nation. Instead, we contribute to our understanding of the complex equation of war by recognizing the lengths to which our government will go to protect the lives of our service members, and ultimately the lives of all our citizens, from the tragedy of war.

Additionally, perhaps, we should take some reassurance in the fact that it is not easy for members of our species to take the lives of our fellow human beings. This is the dominant theme of the book *On Killing*. The hard thing to explain in our civilization is not the one-in-a-million horrible murder but the 99.9 percent of our citizens who will go a lifetime and never once take a human life, or even seriously attempt to do so. Inside most healthy members of our species there appears to be a "hard-wired" resistance to killing one another. Divorce, infidelity, layoffs, and traffic accidents by the untold millions, yet in the face of these provocations it is extraordinarily difficult for most of us to kill one another. Once again, we see this fundamental human dynamic at play in my study of the RPA community, and there may be cause for us to be reassured by the message that is communicated here.

My Worst Day on the Job

To gain a better understanding of how invested RPA pilots are in their mission and how cognitively connected to it they are, I conducted an anonymous survey of 254 RPA crew members and interviewed more than fifty people to tell me about their worst day on the job. Regardless of your opinion of RPA pilots, you can't help but recognize their connectivity to the fight from their own stories.

The book *On Combat* outlines the concept of "no pity party, no macho man." Navigating the balancing act between these two extremes is a critical aspect of dealing with the psychological impact of combat. It is important to note that, in outlining the worst-day experiences of these individuals, I am not laying the foundation for anyone to cop a pity party. Rather, I am staking out one end of the continuum of responses experienced by the RPA community and demonstrating their powerful emotional investment in their mission.

Read their stories and ask yourself this: how much does it matter that these events are witnessed on a live high-definition video screen rather than in person, firsthand? Here are a few of the pilots' responses to my request to tell me about their worst day on the job.

- "The first time you kill. It is the best and worst day. You experience emotions you have never experienced. I remember most of it like it was yesterday. That day fundamentally changed how I thought about war and galvanized why I serve. But the memories come back at strange times, when you least expect it. Therefore, you have to live it over and over again... even when you don't want to." —Lieutenant Colonel, USAF, MQ-9 Reaper Pilot

- "Watching an innocent civilian get decapitated, following the executioner, then, partway through the follow, being asked to drop follow and shift target sets. No other asset was on station and the executioner walked away unscathed." —Captain, USAF, MQ-1 Predator Pilot

- "I was the closest person to a TIC [troops in contact] and gained radio contact with the JTAC [Joint Terminal Attack Controller] on the ground, got eyes on, then watched the JTAC get shot." —Major, USAF, MQ-1 Predator Pilot

- "Watching the Blackwater civilians be strung up and burned in Fallujah." —Staff Sergeant, USMC, RQ-2 Pioneer Pilot

- "I watched some civilians inadvertently get killed by an AH-64 [an Apache helicopter]. I felt like I could have prevented it, but I didn't speak up in time." —Captain, USAF, MQ-9 Reaper Pilot

- "Not being able to engage enemy personnel that we watched execute friendly forces." —Staff Sergeant, U.S. Army, MQ-1C Gray Eagle Pilot

- "Seeing children die during a TIC event where they were caught in the middle of a firefight." —Major, USAF, MQ-1 Predator Pilot

- "I employed a missile on an adult male. It blew off both his legs and he was still alive. Watching him struggle on the ground was difficult to watch." —Captain, USAF, MQ-1 Predator Pilot

- "I watched civilians in an Iraqi city dragged from a building and get executed in the street." —Sergeant, U.S. Army, RQ-7 Shadow Pilot

And some non-pilot responses:

- "Watching the son of the person I just obliterated with a Hellfire missile pick up the pieces of his father. It wasn't the act of killing that I focused on, it was watching the boy's face and interactions with the rest of his family that continue to haunt me." —E-6, USAF, MQ-1/9 Sensor Operator
- "Watching a young child carry half an adult male down the road to get help." —Master Sergeant, USAF, Intelligence Analyst
- "My most traumatic day was observing a kid step on an IED. So being part of a UAV crew exposed me to the brutal reality of war from the comfort of CONUS [the continental U.S.]." —Staff Sergeant, USAF, Intelligence Analyst
- "Cleared for a target, inbound, weapon released. Half a second before impact a minor, approximately nine years old, darts out from a building and into the splash zone." —Staff Sergeant, U.S. Army, MQ-1C Gray Eagle Sensor Operator
- "My worst day on the job was doing convoy overwatch for Marines in Afghanistan. During the convoy's move to infiltrate, they hit an IED and came under heavy fire from Taliban forces. We were flying MQ-1s at the time and could not accurately identify where the fire was coming from due to heavy vegetation cover in the mountains. The JTAC on the ground was watching our video feed while trying to talk to us via radio as well as return fire on Taliban forces. With us searching for the enemy locations, multiple Marines were wounded and one was KIA [killed in action]. Ultimately, we were able to locate fleeing enemies in a nearby compound that was passed to the ground party. I felt *responsible* for the casualties to our Marines because we could not locate the enemies until they retreated to a compound that we could not strike at the time due to CDE [collateral damage estimate]." —Technical Sergeant, USAF, MQ-1/9 Sensor Operator

- "Although I would not do it differently, conducting surveillance on a confirmed HVT [high-value target] and seeing him positively interact with his spouse and children, like a caring father and husband...playing soccer with his son, and ultimately taking a kinetic strike opportunity not far from his home. No collateral damage occurred, however, had full viewing in HD of the family mourning and could not help but think that the son would be the next generation of terrorist due to this event." —Contractor, ISR Tactical Controller for RQ-7/MQ-1/MQ-9

It is evident that there is absolutely a human element to the work being done with RPAs. There were countless other stories recounted to me that reinforce the emotional attachment and cognitive connection of the crew to the mission. The RPA is the tool. It is the crew who is at war, regardless of physical location.

Sensor Operators: Harder and Equally as Important

The sensor operators, predominantly drawn from an occupation field of enlisted personnel, control the onboard sensors to conduct surveillance of potential targets or geographic areas. They also play a crucial role as the last link in the chain when conducting a strike: guiding missiles or bombs onto the target, from launch to impact, a process that takes about thirty seconds, depending on the distance of the target from the aircraft. Following a strike, sensor operators survey the aftermath, conducting what is known as battle damage assessment (BDA), to ascertain whether the strike achieved the desired result.

Joseph, an Air Force Reaper pilot who chose to keep his last name anonymous, had the unique perspective of serving both as a Predator sensor operator and then as a Reaper pilot in his career. He deployed twice to Afghanistan as part of a launch and recovery element and also conducted missions remotely from the United States. Although he deployed twice, all his kills occurred while operating from the U.S. at maximum remote distance. Having performed both as a pilot and a sensor operator, he said that without question the sensor operator job is harder and equally as important, but he felt as though only the pilots received attention in the media. Although he stated that he never suffered any psychological impact from killing with a Predator, what he recounted was interesting and important for understanding the human reaction to killing remotely with RPAs: his physiological responses.

Joseph said that every single time he killed with an RPA he experienced identical physical responses that morphed throughout the course of a strike. When ground forces request an air strike (from either manned or unmanned platforms), the sequence of events and the communications that occur between requester and striker is a standardized, systematic, step-by-step process full of quick verbal exchanges that convey the information necessary to prompt the next step. It is rhythmic and methodical by design. Anyone trained in the application of aviation fires can listen to this exchange on a radio and tell you what is occurring at that moment and what will occur next.

The process facilitates the quick employment of fires when it is needed most and usually begins with a radio call to the crew from a Joint Terminal Attack Controller located on the ground, forward in the country. The verbiage routinely starts with "[Aircraft's call sign], this is [JTAC's call sign], stand by to receive a nine-line." A nine-line is essentially an attack request that entails a description of the target, the target's location, the location of friendly forces in relation to the target, and a recommendation of how the target should be attacked. Immediately upon hearing those words, "Stand by to receive a nine-line," classical conditioning kicked in and Joseph's breathing intensified, his heart began to race, his adrenaline spiked, and he had to steady his hands just to write down the information being passed to him over the radio.

After the information is successfully read back to the requesting unit, the pilot maneuvers the RPA into position for the best weapons-release profile while Joseph searches for the target. Once the target is "captured" in the camera and the airplane is in position to strike, Joseph and his pilot wait for clearance from the JTAC to begin the attack. This correlation of confirming the correct target often occurs over video as the JTAC views the RPA's video feed. The pilot announces that he is set and starts to fly inbound toward the target. Once given the command of "cleared hot" from the JTAC, the attack commences.

At the pilot's command, Joseph fires a continuous laser beam from the aircraft to mark the target for the laser-guided munition. The seeker head attached to a laser-guided missile or bomb then looks for the laser energy on the target. The pilot initiates the strike with a call of "three, two, one, rifle," which is military terminology for missile launch. Upon launching a missile, Joseph said it always feels as though his heart stops momentarily and his adrenaline spikes again.

For the remainder of the strike, success rests entirely on Joseph to guide the missile onto the target by keeping the laser beam pointed at it. As the missile approaches the target, the pilot counts down backward from five until the time of impact. Upon hearing the countdown, Joseph's gaze at the target intensifies, and he oftentimes finds himself holding his breath and experiencing tunnel vision of the target so intense he can't hear or see anything else. Upon impact of the missile, the screen flashes and blacks out for an instant. Joseph's role then shifts to evaluating the effectiveness of the strike and looking for anyone fleeing the impact site—referred to as "squirters" in military parlance.

During one mission, while providing overhead protection for friendly troops fighting ISIS in Iraq, Joseph watched as a "technical" vehicle (a Toyota truck with a large machine gun mounted on its back) maneuvered around the city, fired at friendly forces, and then ducked into a garage. This process repeated three times until the ground troops finally gave Joseph and his pilot the approval to strike the vehicle. Due to the urban canyon nature of the city, the missile had to travel down the middle of a very narrow road the length of several city blocks before it reached the technical. Needless to say, it was a very complex shot from nine kilometers away that ended with a destroyed technical vehicle and dead enemy fighters who had been firing at friendly troops all day.

Joseph described another complex emotion that swept in as he and the pilot exchanged an ardent high five after the strike. Thinking about that moment and how he appeared to be celebrating the death of an enemy combatant, Joseph paused and searched for words before he continued. "The feeling of elation or joy post-strike came from solving a challenging problem, not killing someone."

To recap Joseph's sequence of physiological responses to killing remotely: heavy breathing, elevated heart rate, adrenaline rush, shaky hands, disruption of breath, tunnel vision, diminished sound, joy, and elation. As discussed in *On Combat*, these physiological responses are similar to those experienced by a warrior in a face-to-face shooting incident. That these responses transfer to someone killing from maximum range demonstrates two important points: the mind does not distinguish the event mediated through a screen as happening seven thousand miles away and, clearly, there can be an emotional and physical reaction to killing, regardless of distance.

Intelligence Analysts: An Intense Connection to the Mission

Although RPA strikes garner a great deal of attention, RPAs are predominantly used as an intelligence-gathering platform. An airborne camera that flies for a long time relatively undetected over an area or person of interest is well suited to figure out what the hell is going on down there. Therefore, conducting ISR (intelligence, surveillance, and reconnaissance) dominates the airborne time for any RPA. Missions are often pervaded with extreme, personally fulfilling, and excruciatingly long expanses of time for the crew, spent staring at rocks, dirt, roads, buildings, friendly forces, or people going about their normal, everyday lives; just waiting to catch a glimpse of something abnormal, out of place, or a threat to friendly forces. Combat is often described as long periods of boredom interrupted by brief periods of excitement. This accurately describes RPA flights.

The Air Force alone flies more than 550,000 RPA hours every year to support the global demand from geographic combatant commanders. Every second captured on video is analyzed and processed by humans who specialize in intelligence. It's the equivalent of one hour of RPA video uploaded every minute of every day for the entire year. Analyzing that amount of video to make sense of it is beyond a major undertaking. As a matter of fact, it takes an entire intelligence enterprise to do it in the Air Force.

The Distributed Common Ground System, as that intelligence enterprise is known, consists of twenty-seven distributed sites around the globe, networked to the manned and unmanned ISR assets they support. Each location consists of a mixture of active, reserve, and Air National Guard intelligence analysts who provide intelligence support to real-time tactical missions. Analysts assist with pre-mission intelligence planning, intelligence support during the mission, and development of post-mission summaries.

In short, for every Air Force operational RPA flight, there is a team of analysts watching the live video feed and providing feedback to the RPA crew, particularly when it comes to positive identification of a target. This information is then conveyed to the crew through an intelligence liaison embedded with the crew called a mission intelligence coordinator, or MIC. The ability of RPA crews to talk directly with a network of intelligence analysts during the flight provides a distinct advantage over manned aircraft.

The other branches of the U.S. armed forces don't have an extensive distributed network of intelligence analysts but, rather, have the analysts sit directly with the RPA crew during the mission. They, however, are a fully integral part of what occurs during the flights and, based on their role in deciding whether criteria are sufficient to strike, share in the decision-making process and the accountability.

What I just described is a distributed, digitally connected crew-served weapon. As discussed in *On Killing*, we know that employing crew-served weapons is an effective means for overcoming individuals' resistance to killing. This "diffusion of responsibility" is extremely important and powerful, and it spreads the burden, culpability, and responsibility, and allows a certain level of anonymity in the crew. More on this later when I discuss RPAs and group absolution. But for now, understand that regardless of whether you're the pilot flying the RPA or the analyst determining the validity of the target, you have an intense connection to the mission.

One experienced Air Force intelligence analyst described to us her work with RPAs:

In the beginning, I thought this would seem very remote or "video-gamish," that this would be easy. But it's not. There is still a human factor there. I saw it in other airmen, they were changed by their experiences involving killing. One airman I knew was so troubled by the job that they decided to quit. At the time, there was no help available, and the sentiment in the community was "if you can't handle it, get the hell out of here" and they were shamed out of the community. Initially, nobody would talk about killing. There was command pressure to keep everyone focused on executing the mission. The prevailing attitude was that if everyone's talking to the chaplain, no one's executing the mission. But today mental health support has changed for the better.

Eventually, I became numb to all of the killing and the things I had seen. It affected my sleep. I started second-guessing my training; was that the right target? The details of what I witnessed were extremely intimate and humanizing. I saw people's distinctive gait of their walk, their mannerisms, kids playing soccer, people feeding their goats, couples having sex on the roof of their home, people taking a crap outdoors, guys having sex with animals, and guys planting bombs on the side of the road. Sometimes it was obvious who the bad guys were, and when I saw them shooting at friendlies it made the decision to kill them easier. It was always easier to kill when you knew you were protecting friendly forces than when you were going after a specific person. That changed when I was at a squadron that hunted specific people.

I think there is a misconception out there that the intelligence community isn't involved in the fight, but I was responsible for the death of a surprising number of people. I got to know my targets better than my own roommates. I knew everything about them. I could tell you if the target had lost or gained weight in the last few months, the names of their children, where they lived and how they spent their days. What really added to the humanity factor was when I got a picture of the person we were targeting. One target, due to the Russian influence in Afghanistan, surprisingly to me had blond hair and blue eyes and looked like my neighbor back home. It was really hard to kill him.

What countered all that humanity I observed of the target was that as an intelligence analyst I was also privy to all of the nefarious activity that the targets were responsible for. That knowledge made killing a high-value individual [HVI] more personally satisfying than killing a farmer who was paid to pick up a gun and shoot at Americans. That farmer was just trying to provide for their family.

What I found critical to helping the crews was sharing the "why" each HVI was being targeted. Sharing information built trust between the crew and intelligence, and trust has to be there or it sows a seed of doubt in the crew's mind about the legitimacy of the target. Otherwise the crew is left with the question of why did I just kill that person. I personally witnessed this frustration from a crew once when a pilot was screaming on the phone with the intelligence enterprise and refused to order the strike until an intelligence analyst provided the "why" behind the strike. It was extremely emotional. I enjoyed my time working with RPAs, but it wasn't without a price. At one point, I showed up on an ISIS hit list for the work I did. It was very real and very scary.

It isn't difficult to understand why two-thirds (twenty-eight of forty-two) of the intelligence analysts I surveyed said they had experienced a negative response to killing in their career. Half of them (twenty-one) had experienced multiple negative responses to killing. Six analysts (14 percent) attributed their negative responses to feeling responsible for causing civilian casualties on the battlefield without ever stepping foot in that country. Even if you are not the one pulling the trigger, the emotional connection to the kills is there for intelligence analysts in large part because of the humanity they witness in the targets. As discussed in *On Killing*:

> When you have cause to identify with your victim (that is, you see him participate in some act that emphasizes his humanity, such as urinating, eating, or smoking) it is much harder to kill him, and there is less satisfaction associated with the kill, even if the victim represents a direct threat to you and your comrades at the time you kill him.

But as the analyst quoted above mentioned, that humanity was countered by the knowledge of the nefarious activity the targets were responsible for. This important factor is discussed in more detail in a later chapter on target attractiveness.

By now it should be apparent how emotionally connected to missions our RPA warriors are, regardless of their position in the crew and regardless of where that crew is located in the world in relation to RPAs.

Chapter Five

The Missions

Although at this point our thoughts about what the future will be like when robots and humans co-exist is speculative, one thing is fairly certain—humans will no longer be employed doing jobs that robots can do safer, faster, more accurately, and less expensively.

—Bernard Marr, futurist

Dull, Dirty, Dangerous, and Difficult: The RPA Sweet Spot

Some working-class, blue-collar jobs in the United States have been described as dirty, dangerous, and demeaning, colloquially referred to as the 3Ds. Historically these jobs involved hazard to a worker's health, even possible death, and have been predominantly filled by immigrant workers seeking better pay, opportunities, and quality of life than they could achieve in their country of origin. This was evident during the great European migration to the U.S. between 1880 and 1920 that initially co-incided with the Gilded Age in America. Large populations of immigrant workers found employment in manufacturing, mining, or construction fields in a less-than-ideal safety environment. This trend continued over time as the source of the immigrant population shifted from Europe to Asia to Latin America.

With the rise of robotics in the twenty-first century, the labeling of these 3D jobs has been supplanted by a new designation that describes a job ideally suited for a robot: 4Ds—dull, dirty, dangerous, and difficult. As technology has matured, it has become more economical to buy and "employ" robots, which began replacing working-class labor in traditional 3D occupations. And this trend will continue, according to the International Federation of Robotics, with an estimated additional 1.7 million robots joining the workforce by 2020.

We see this playing out in the commercial sector, where robots are used to move items around warehouses (dull); inspect the structural integrity of bridges, power lines, and windmills (dangerous); look for cracks and leaks in sewer pipes (dirty); and assist doctors in surgeries (difficult). We also find robots performing 4D missions in the military: conducting surveillance for long periods of time (dull); defusing makeshift bombs (dangerous); inspecting an area for nuclear, biological, or chemical contaminants (dirty); and precisely killing an individual at an appointed time from half a world away (difficult).

Robots performing military missions in the early twenty-first century, however, are not fully autonomous and therefore require varying degrees of human input to influence the execution of their algorithms. Until robots are fully autonomous on the battlefield, it's best to think of humans as executing missions *with the aid of* a robot, rather than the robot executing the mission independent of the human. More important, since humans are still very much central to this intricate equation, we must investigate this man–machine relationship. Supporting this inquiry is the finding that there is very limited research on how fighting via a robot in combat, especially when using a robot to kill, triggers psychological responses. It's fairly uncharted territory. To understand the human experience of working through a robot, we must look at the missions they perform.

Watching Paint Dry: Intelligence, Surveillance, and Reconnaissance

Military RPAs are ideally suited for 4D work but are primarily used to collect information through video imagery that is analyzed and turned into intelligence. For simplicity, I'll classify anything that doesn't involve attacking a target as an intelligence, surveillance, and reconnaissance mission (better known as ISR). ISR missions include watching over convoys, patrols, or other forces during their movement, searching a specific area or route for enemy equipment or fighters, screening an area to report detection of enemy personnel or equipment, collecting information about a geographic area or building that will be raided or attacked in the future, conducting search and rescue of friendly forces, conducting surveillance of the local population within a specific geographic area to determine a baseline of "normal" activity (known as pattern of life), and tracking high-value individuals.

The primary payload used for ISR missions is an electro-optical/infrared camera, but additional payloads such as a synthetic aperture radar (SAR), a ground moving target indicator (GMTI), or a signals collection payload can provide intelligence to supplement the full-motion video or serve as a stand-alone intelligence source. Either way, an RPA employed in an ISR mission is looking for somebody, listening for somebody, or both. Unless the situation "goes kinetic," meaning it evolves into an event necessitating lethal force.

A common trend in the use of RPAs among all groups is to perform what's referred to in the military as "battle damage assessment" or "bomb hit assessment," meaning after a target is struck by an RPA or another weapon, the RPA crew observes and reports the aftermath and effectiveness of the strike. It is during this part of the mission that RPA crews routinely witness the carnage of war. How those exposure events affect RPA crews will be discussed in Section II.

Pattern of Life: Determining a Baseline

Until artificial intelligence technology has matured enough to handle the dull portions of RPA missions, that burden rests with humans. One such extremely dull activity is the pattern of life (POL) mission, which usually involves watching a certain portion of a population or a single person over a long period of time to establish a "baseline" of their normal activity, therefore enabling you—in theory—to identify when abnormal behavior occurs or, in the case of a high-value individual, when the time is right to kill them. And nothing does it better than RPAs.

POL missions emerged as a by-product of fighting long counter-insurgency wars in which the enemy typically hides among the civilian population, and to kill the enemy you first have to find him. The challenge is that establishing a good pattern of life takes time to develop. Which means that the crews oftentimes find themselves looking at the same area, same building, or same person for days, weeks, months, or even years. It is incredibly boring. One Reaper sensor operator's comments said it all when he told us, "POL is boring as fuck." As much as that statement may be true, POL missions in which a crew watches an HVI for a long period of time are also incredibly intimate. This intimacy derives from the fact that POL tends to reveal the humanity of the target. And observing the humanity of an individual, as we know, definitely doesn't make killing them any easier, psychologically or emotionally.

POL missions can become deeply voyeuristic. One young Reaper sensor operator told us how he found himself forming this strange bond with his target. Over the course of six months he would come to work and spend eight hours a day watching this HVI go about his life, playing with the kids before school, walking the kids to school, traveling to meetings, going to the market, etc. . . . As the sensor operator put it, "I spent more time with this guy than I did with my family. I knew everything about him: his face, his family, his mannerisms, and his gait. I could pick him out in a crowd full of people." This routine continued for six months, waiting for the HVI to venture off into what's referred to as "green terrain," meaning free of the possibility of collateral damage, thereby presenting an opportunity for him to be killed.

Once, the sensor operator took a routine day off from work. When he returned, he discovered that another crew had killed his HVI in his absence. He felt devastated, as though something had been stolen from him. He was angry. That was his kill, and someone took it from him! Why did he feel this way? Why did it matter who got to kill this bad actor?

The sensor operator invested six months of his life developing POL with an assumption that the mission would end with him striking the target. Professionals want to finish a task through to completion and that opportunity was taken from him. Over the course of six months, the sensor operator developed a one-sided intimacy with this individual. The HVI's death in the sensor operator's absence produced a strange sense of loss that left the situation unresolved in his mind. But, depriving the warrior of the kill is an effective technique to break the one-way emotional bond formed between killer and victim.

Killer Missions

The military refers to air strike missions as offensive air support (OAS), with two subcategories of close air support (CAS) and deep air support (DAS); the difference being that CAS (pronounced "kass") is conducted in close proximity to friendly forces and requires detailed coordination and integration with ground forces for approval to strike. A request for a CAS mission can be either preplanned (planned before the flight) or immediate (reactive during the flight). CAS missions are either scheduled (the aircraft shows up at a certain place and time) or on call (on standby either on the ground or in the air). The majority of RPA strikes are immediate, on call, or dynamic strikes; meaning an armed RPA is airborne waiting for a situation to present itself that requires someone to be killed or something to be destroyed. The most common and easiest example is when friendly troops are ambushed by the enemy. This is known as a troops-in-contact (TIC—pronounced "tick") situation.

TICs provide the greatest cognitive and empathetic connection to the battle for RPA crews, and are subsequently one of the most emotional and distressing missions when things go awry. This mental connection hinges on one integral factor during support to TIC missions: friendly troops fighting for their lives. Since TICs are dynamic missions requiring immediate assistance, most times when an RPA shows up to support a TIC, they are directed to quickly find and destroy enemy forces that are attacking friendly forces. As would be expected, when you arrive on scene in this situation, the friendly forces being shot at by the enemy express a wide array of emotions over the radio such as fear, anger, frustration, and impatience that can be felt and understood on the other end of the radio regardless of which crew position you occupy. Often, troops' lives literally depend on the integration and assistance of the RPA called to help them fight through the dangerous situation they are in. And you can hear the stress in their voice. When that situation goes badly, when you fail to protect your fellow warriors on the ground, no doubt you are affected by it whether thirty thousand feet overhead in a jet or seven thousand miles away in a detached cockpit on the ground.

When the enemy is hiding among the civilian populace in an urban environment, most strikes require detailed coordination and integration with a ground force commander and are therefore close air support in nature. I'll talk about this more in the chapter on demands of authority, but for now understand that the preponderance of RPA strike missions requires the ground force commander to approve it; in other words, RPA crews aren't initiating or approving their own strikes as they could if employed in a deep mission.

Another main mission that involves striking a target is hunting for high-value individuals. HVIs range from a low-level combatant or finan-cier up to the leader of a violent extremist organization, such as Osama bin Laden. HVIs are individuals on an approved target list and their nomination to that list is based on some nefarious activity they have taken part in or a military advantage gained by targeting them.

The approval required to strike an HVI also varies but can reach as high as the commander in chief, the president. The hunt for an HVI can be a long and tedious process spanning many different crews and aircraft and can last for days, weeks, months, or even years, waiting for the right opportunity to strike the target while eliminating or mitigating civilian casualties. HVIs often are aware of this fact and use civilians and even family members as shields to prevent themselves from being targeted. The importance of the HVI usually determines the acceptable level of risk associated with the mission and the approval authority. One Reaper pilot I interviewed recalled flying a mission on Thanksgiving hunting for an extremely important HVI, with the instructions that once you are ready to strike let the operations center know, the phone next to you will ring, it will be the president of the United States, and he will authorize the strike.

Another Reaper pilot recalled how long and exhaustive some of these hunts for high-value individuals can be:

One shot, we spent years tracking a high-valued target. We had lost him once or twice, but finally had him again. We knew he needed to go, but it still took us months to attack because our supporting unit had a strict zero collateral damage rule. We watched the man for months do the same routine, go to the grocery store, come home, go to a meeting with his cohorts, come home. But he always brought his kid with him because he knew we wouldn't strike him then. Four to eight unmanned aircraft watched this man 24/7 for months, until one day he finally slipped up and took a quick trip without his kid and we were ready immediately to strike and we killed him.

Even unarmed RPAs can participate in strike missions in a support capacity. Oftentimes this indirect role involves locating a target, passing that information off to another asset to conduct the strike, and then providing the post-strike assessment. Some unarmed RPAs carry laser designators and can minimize the striking aircraft's exposure to the enemy by lazing in guided munitions to the target for them. Even armed RPAs are used in this manner due to their long loiter time and ability to develop a situation and then call in jets or helicopters for the attack. Unarmed RPAs can also be used to call fire support missions for artillery, rockets, or naval gunfire.

As an example, a crew flying an RPA during an ISR mission may discover a target. With its onboard sensors, the precise location of the target is determined and transmitted to a fire support system, say a High-Mobility Artillery Rocket System (HIMARS). Once approval to fire on the target is given by a fire support coordinator, the HIMARS crew then launches a GPS-guided precision rocket at the exact grid where the airborne, unmanned forward observer found the target. These precision-guided rockets are as accurate as GPS-guided bombs but have less explosive yield and are therefore a preferred weapon in situations where collateral damage adjacent to the target, such as in a city, is a big concern. Regardless of an RPA's size or armament, it can be used directly or indirectly to strike targets. This is one of the reasons my research included all sizes of RPAs and all services that fly them, instead of just armed RPAs flown by the Air Force, so that I could compare and contrast how the crews responded to using different types of RPA to kill.

Robot Maverick: The New Top Gun?

An emerging mission for RPAs, and one they are well suited for, is in a traditional air-to-air fighter role. While the technology to fulfill this mission is still in its infancy, it isn't hard to see why an armed, unmanned platform, digitally connected to early warning and fire-control radars and networked to other aircraft to determine the best platform to shoot from, would be well matched for this mission. Another reason RPAs might be effective in air-to-air missions is that high-G maneuvers during "dog-fights" are limited by two things: stress on the aircraft and stress on the pilot. If you remove the pink squishy thing known as the pilot, the aircraft can perform more high-G maneuvers since the pilot will normally lose consciousness before the plane becomes overstressed. In a dogfight against an enemy pilot, that could be just the advantage needed to win the duel. So we are moving toward having unmanned fighter aircraft. But it is a slow stroll down a path with many obstacles.

The X-47B, an Unmanned Combat Air System-Demonstrator (basically a fighter RPA) developed for the U.S. Navy by Northrop Grumman, is a stealth RPA capable of taking off and landing on an aircraft carrier and carrying 85 percent of the ordnance of an F-35 out to a range of 2,400 miles. In 2015, the success of X-47B tests prodded Ray Mabus, then the secretary of the Navy, to state that the joint strike fighter "almost certainly will be the last manned strike fighter aircraft the Department of the Navy will ever buy or fly." As of the present day, it's still to be determined whether that statement will ring true.

The X-47B was only a demonstrator aircraft and didn't make the transition to a funded program of record where the Navy would actually buy and employ them. Instead, the U.S. Navy focused on getting an RPA capable of in-flight refueling of the manned airplanes launched from aircraft carriers. Currently, Super Hornet jets launched from aircraft carriers spend 20–30 percent of their time refueling other jets rather than in their primary role as a fighter or attack aircraft. In-flight refueling is a dull mission and not a good use of a manned fighter aircraft's limited time, both in the air and over its life span, but it is a dull mission ideal for an RPA.

In August 2018, the Navy awarded a contract to Boeing to build four MQ-25A Stingray aerial refueling RPAs. Initially, the Stingray was going to be a stealth aircraft that, in addition to aerial refueling, could also serve as an attack aircraft dropping bombs. However, before it was built, the requirement for the Stingray to be a stealth aircraft was eliminated, almost guaranteeing that it would not be a threat in replacing manned platforms in the attack role for the Navy anytime in the future. Having an aerial refueling RPA will, however, allow manned fighter aircraft to accomplish more of their traditional fighter and attack missions. The future of fighter RPAs launching from an aircraft carrier remains unclear for the Navy, but one thing is certain: There are a lot more obstacles in the path to replace humans in the cockpit.

The Air Force has been experimenting with RPAs in an air-to-air capacity for years. For example, despite the fact that the short-range, antiaircraft Stinger missile has been in the U.S. inventory since 1986, the only time it has ever been fired in combat by U.S. forces was off a

Predator. Walter Boyne told of the encounter in an *Air Force Magazine* article titled "How the Predator Grew Teeth":

> ...some Predators were armed with the AIM-92 Stinger missile, to defend themselves against Iraqi fighters. Getting the Stinger certified on the Predator took only 91 days.
>
> On Dec. 23, 2002—less than three months before Operation Iraqi Freedom began—a Stinger-armed Predator was performing reconnaissance over a no-fly zone when an Iraqi MiG-25 turned in to attack. The Predator fired at the MiG-25, and the TV imagery showed the smoke trails of the two missiles crossing in midair. Unfortunately, the MiG's missile downed the Predator, but the Iraqi Air Force apparently drew the conclusion the US would have wanted them to: that there was no future in combating Stinger-armed Predators. There were no further attacks against the UAVs.

That is the benefit of using an unmanned aircraft in an air-to-air mission with a high likelihood of getting shot down. No one dies. Although crassly put, Richard A. Clarke, President George W. Bush's chief counterterrorism adviser, emphasized this point when he said, "You know, if the Predator gets shot down, the pilot goes home and fucks his wife." The sentiment that RPA pilots are completely detached from all aspects of the mission by being physically removed from the cockpit is, as we can see, pervasive and far reaching.

In March 2018, the Air Force announced that it was going to arm the Predator's replacement, the MQ-9 Reaper, with the AIM-9X Sidewinder, a heat-seeking, air-to-air missile. This seems like an odd venture and hopefully won't reach the same fate as the Predator/Stinger combo. The Reaper isn't optimally designed to serve as a fighter aircraft. Without diving too deep into a nerdy aerodynamics discussion, simply put, aircraft designed for longer endurance flight such as the Predator, Reaper, or a glider were not designed for maneuverability based on the aspect ratio of the wing.

Higher aspect ratio wings, such as gliders, have a greater lift-to-drag ratio and are therefore more fuel efficient and stay aloft for longer periods of time. Airplanes with a low-aspect ratio, with shorter and wider wings, are faster and more maneuverable. If you have ever seen an F-14 Tomcat (and I know you have because it was the jet that Maverick flew in *Top Gun*), it is a prime example of how the aspect ratio affects an aircraft. The Tomcat had what's referred to as a variable-sweep wing, which means that it can change its aspect ratio during flight to be more efficient or effective to suit its purpose for the mission.

In contrast, the Reaper is designed for endurance, so why consider it for an air-to-air role? One potential employment option could be an airborne phalanx of Reapers serving as "missile trucks" defending an area with beyond-visual-line-of-sight missiles. This would seem to be the ideal employment method of arming the Reaper with the AIM-9X or potentially a radar-guided missile such as the AIM-120. Whatever the U.S. ultimately decides to do in the air-to-air realm, it is a high probability that the sixth-generation fighters that replace the fifth-generation F-35 Lightning will include unmanned platforms. When that happens, when we employ RPAs in an air-to-air mission and their engagements occur beyond visual line of sight, targeting objects like blips on a radar screen rather than an individual person on the ground, the killing involved in that mission will become less personal and less graphic for the RPA crew.

Turning Chicken Shit into Chicken Salad: The Need for Better Tasking

Over the last twenty years, the military has treated RPAs like belt-fed ammunition, abundant and inexpensive. This has resulted in less-than-optimal mission tasking and a burnt-out force. Tasking that occurs through either the intelligence or operations chain usually elicits a food fight over who owns the process of directing RPAs. The tasking that comes down to the RPA unit is often too watered down to be useful (e.g., "scan the roads and look for something out of place"), too specific to be executable ("find an individual in a crowd"), or outside the scope of capabilities of the platform ("scan the city for enemy combatants").

The lack of understanding of RPA capabilities by leaders contributes to the poor tasking described above. The net result is wasted hours, wasted dollars, and wasted resources spent staring at something that provides little military value. If it sounds frustrating, it is. But imagine you had the responsibility of providing meaningful tasking for an RPA over a twenty-four-hour period. Odds are you would run out of decent tasks after a few hours and just resort to having crews "watch" something.

The nonchalant tasking of these valuable assets results in lots of missions where RPA crews "look for hotspots on the side of the road," in the desert, in Iraq. Everything is hot in the desert, in Iraq. That tasking may actually be sufficient if you can use automation and artificial intelligence to fly the RPA and monitor the video to prompt the user when something of interest appears. But the technology isn't there yet. So a fair portion of the hours spent flying ISR missions yields little return on investment.

Christopher Drew, in a *New York Times* article titled "Drones Are Weapons of Choice in Fighting Qaeda," quoted one RPA operator about the countless hours spent doing mindless tasks: "We spend 70 to 80% of our time doing this, just scanning roads." Drew also made the point that during a two-year span from 2007 to 2008 of RPAs supporting the wars in Iraq and Afghanistan, out of 10,499 missions flown by RPAs, only 244 of those missions resulted in a strike. That means 98 percent of all missions flown during that period were strictly ISR missions, with a large majority of them spent "scanning the road."

Unfortunately, compounding this issue is the process the U.S. military uses to determine how many RPA flights each combatant commander (the general/admiral who leads U.S. military operations for a geographic portion of the world) receives. RPA flights are viewed as use or lose, meaning if a combatant commander doesn't use all the RPA hours that have been allotted to him, he runs the risk of losing them to another combatant commander. And no commander ever says he has too many ISR hours! But if he did, it might provide some desperately needed relief to an overworked and overstressed community.

But using an RPA is cheaper and safer than using a manned airplane to fly what's called a nontraditional ISR (NT-ISR) mission. Which, unfortunately, is what the U.S. did for most of the war in Iraq while it built up RPA capacity to take over these missions, literally burning through the limited useful jet hours those platforms had to look for low-level insurgents planting IEDs. Colonel Kevin "Astro" Murray, an F/A-18 Hornet pilot and former commander of a Marine Corps RPA squadron, described this misuse of assets in a *Marine Corps Gazette* article titled "Marine Aviation Readiness":

> As the force transitioned from major combat operations (MCO) to irregular conflict, we chose to leverage what we had, vice asking the more difficult question of "what is the best tool for the job?" F/A-18 and AV-8B missions began to be flown based upon electro-optical and infrared full-motion video cameras, and thousands of sorties were flown where tactical jet aircraft never dropped a single weapon. . . .
>
> The results of these decisions, over the past 15 years, are readily apparent. Aircraft that were meant to be used for deterrence against near-peer adversaries and for fighting in MCO have found their flight life used up in the NT-ISR role fighting insurgents. Jet aircraft meant to fly fast to survive and strike robust military targets burned their life away "in the overhead" as surveillance and reconnaissance assets.

While ISR is not the most effective tasking for the people flying RPAs currently, it is better and cheaper than any of the alternatives. And better and cheaper—that is the inherent goal of all technology, even if the modus operandi is dull, dirty, dangerous, and difficult. Despite the 4Ds, there is of course unequivocal benefit gained in protecting friendly forces and providing precision strikes.

Unfortunately, RPA crews routinely spend the preponderance of their time conducting ISR or hunting HVIs. Over time, this exacts a toll. And that toll must be paid by RPA crews, as will be discussed in later chapters. For now, let's shift our focus to the specific ways RPAs are employed to execute their assigned missions.

Chapter Six

The Methods

Necessity is the mother of invention.

— Richard Franck, *Northern Memoirs*, 1658

Born of Necessity: Remote Split Operations

Revolutionary inventions are often born of necessity. The tin can, ubiquitous today, is a prime example and, like many inventions, it started with a problem in the military. Before the days of canned food, a large army of warriors on the march would quickly run out of fresh food or the opportunity to kill or forage for food in quantities sufficient to feed itself. For an army, it's a pretty simple concept: no food, no march. Napoleon Bonaparte, considered to be one of the greatest military minds in history, understood this well and was once quoted as saying, "An army marches on its stomach."

To solve the problem of keeping his army supplied with food and on the march over long distances, Napoleon challenged his countrymen in 1795 to devise a method of preserving food so that it could easily be transported. The winner, Nicolas Appert, developed a rudimentary method of canning food in a glass container with an airtight seal and in 1810 received the prize of twelve thousand francs for his invention. In that same year, Englishman Peter Durand received a patent for a similar concept from King George III. Durand, however, used the less fragile tinplate instead of glass. This patent was sold two years later, and the process of refining tin cans continued over many iterations, including a migration to the United States, where canned rations played a significant role in keeping the armies fed during the Civil War. It was then that another invention born of necessity finally appeared fifty-five years after the tin can was patented: the tin-can opener.

How armed RPAs came to be flown from half a world away is another story of invention born of military necessity. This genesis is worth spending a little time explaining because it ultimately shaped the Air Force's remotely piloted aircraft program and gave birth to challenges never before faced by a military: killing from maximum distance, then commuting home in time for dinner with the family. Richard Whittle describes in his book *Predator* how remote split operations (RSO) came into existence. What follows is a summary of Whittle's account combined with details I gathered through interviews.

If you recall from the introduction, an unarmed Predator was deployed to Uzbekistan in 2000 to search for Osama bin Laden in the neighboring country of Afghanistan. Prior to 9/11 Uzbekistan wanted to keep the Predator's presence covert, so only a small crew necessary to launch and recover the planes was deployed there. In the previous deployment of the Predator to Bosnia, the ground control station had been co-located with the launch and recovery site, which limited the range of the Predator to the distance the ground antennas at the GCS could communicate with the aircraft. In the mountainous terrain of Uzbekistan and Afghanistan, that limitation of range just wouldn't do.

So out of necessity a plan was devised to control the Predator over a satellite radio. The only problem was the ground control station needed to be located somewhere in Europe or Africa. Without geeking out too much on the explanation, imagine a flashlight shining on a beach ball.

The light hitting the ball's surface represents a satellite's physical area of coverage on the Earth. You can use a satellite to communicate anywhere between two points in the lighted area depending on signal strength of the transmitting antenna and reception capability of the receiving antenna. Unfortunately, in this case, the United States was on the dark side of the beach ball. So the place within the light chosen to support this new Predator capability was a secret compound inside an American facility on Ramstein Air Base in Germany. Secret in the fact that the Germans were not notified of this new development at the time.

The plan worked well in its nascent stage when the Predator was unarmed. However, by September 2001, the Predator was armed, which presented the additional political considerations and challenges of flying an armed RPA from Germany.

As if that weren't a large enough obstacle to overcome, this first cohort of Predators flown over Afghanistan was piloted by the Air Force but was tasked by the CIA. Naturally, the CIA wanted to have the GCS of the Predator close to its intelligence infrastructure, and what better place than the parking lot of their headquarters in Langley, Virginia!

The challenge of how to control the plane from this dark side of the beach ball remained. Fortunately, a scientist involved in the Predator program since inception, who chose to remain anonymous, had an idea for how to solve the problem. To get the signal from Germany to the United States it would be relayed over fiber optic cable that ran beneath the ocean. So the path to control the airplane remotely would look something like this: The pilot in Langley makes an input into the GCS to initiate a left-hand turn. That command is sent near instantaneously over fiber optic cable to Germany, where it is transferred to a satellite radio and transmitted to the aircraft in flight over Afghanistan. The aircraft receives the signal, executes the command, and relays back to the GCS through satellite and fiber that the command was executed. The pilot receives visible feedback on the heads-up display (HUD) indicating the aircraft is in a left-hand turn. And remote split operations were born, of necessity.

Due to the overall distance involved and the time required to relay a signal among several systems, this process is no longer instantaneous. The time differential between initiation of command to feedback registering on the pilot's screen is referred to as latency. It's said in aviation that you don't want to be behind the airplane; in other words, playing catch up in your decisions as the plane continues to fly. When piloting an RPA remotely from the U.S. you are always behind the airplane, 1.8 seconds behind to be precise. Unfortunately, this latency is both inherent and unavoidable and will remain a system limitation until better technology or a different method is devised. Fortunately, for aircrew flying RSOs, the latency is the norm. Irrespective, this concept of flying RPAs remotely has existed since its inception in 2001 as the employment method of choice. In fact, all sixty-five of the combat air patrols (each a twenty-four-hour mission in a geographic area) the Air Force flies worldwide are flown in this manner from the United States. RSO has been extremely effective from an operational viewpoint, but as Whittle described in *Predator* its significance was initially unrecognized:

> At the time, even Werner [a pseudonym for the anonymous scientist] failed to grasp the technological revolution that would follow if he found a way to make remote split operations of the Hellfire Predator work. For the first time in history it would be possible to target and kill an enemy much the way a sniper does—from ambush, and with precision—but from the other side of the world. Science fiction would become science fact.

Remote split operations consist of three major elements: the launch and recovery element (LRE), the mission control element (MCE), and the intelligence network, known as the distributed common ground system (DCGS). I discussed the DCGS in the chapter on people, so you'll recall it is a global, digitally connected system of intelligence support for each USAF RPA mission.

The LRE can be considered the "pit crew" for the RPA. This crew conducts all the maintenance of the aircraft, fueling of the aircraft, preparing the payloads for the mission, loading the ordnance, conducting pre-flight and post-flight inspections, and performing the launch and recovery of the aircraft. It is too risky to fly the entire mission using the satellite control link with 1.8 seconds of latency. Therefore, flying within the terminal environment of an airfield to take off and land is done directly over the line-of-sight control link to the airplane from the GCS at the LRE, removing the 1.8 seconds of latency between control inputs and actions of the airplane. It should be fairly obvious why in this environment you want the actions of the airplane to instantaneously match the control inputs of the pilot. If it isn't obvious, think about how far you can travel in a car going sixty miles per hour in two seconds; that's over half a football field.

In theory, launch and recovery elements are no different in their training or capability from the mission control elements; both crews are manned by qualified pilots and sensor operators. The difference is their physical location during the mission. Every RPA flying over a foreign country is prepared and launched by Air Force personnel forward deployed somewhere in the world. The LRE is just as integral as the MCE, but it is behind-the-scenes work that receives little fanfare and attention. The LRE's aircrew spend time forward in the country where the RPAs are based and then rotate back to the United States and spend time flying as part of an MCE. As such there is a large portion of RPA personnel in the Air Force who have been deployed in harm's way. All the people I interviewed who had served on an LRE relished the experience and said it gave them a better appreciation and perspective of the war they were supporting from stateside. It made them feel more connected to the war.

One LRE Reaper pilot I interviewed described a complex attack on the base where he was stationed in Afghanistan. Once the base came under attack, he coordinated to get a Reaper passed back to him from the MCE to help locate and engage the enemy personnel assaulting the base. Given this capability, it is quite common for RPAs to be incorporated into the defense plan and surveillance plan at the base where they are located. This is just common sense; obviously, it is in the self-interest of the LRE crew to look at the surrounding environment every time they launch and recover an RPA, similar to a convoy leaving the base.

Once the airplane is airborne and positioned in a predetermined piece of sky by the LRE, it is handed off to the MCE through a process called a control station transfer. Think of a control station transfer as similar to the way your cellular phone switches between towers as you drive along a road. Your mobile device recognizes it's receiving a signal from a new location, but its performance is unaffected by the switch in towers and the switch is invisible to you. The only real difference between this analogy and a control station transfer is that—as you would expect—there is a physical electronic approval process included between the two control stations when the transfer occurs.

If we divided an RPA flight into two segments, the cost of doing business (administrative portion) and doing business (tactical portion), we would see that the LRE handles the administrative portion and the MCE handles the tactical portion of the flight. The division is also evident in the actual flight time each element performs. Effectively, the LRE's flight time is marginal (1 percent) compared to the MCE (99 percent), so in a twenty-hour flight the LRE may get a total of twenty-four minutes for the takeoff and landing while the MCE gets the remaining nineteen hours and thirty-six minutes. As great as the lion's share might sound, if you are part of an MCE crew it can be a bit surreal to fly all day but never take off or land the airplane; your flight starts with an airborne RPA and ends with an airborne RPA. But, as alluded to earlier, the MCE gets all the fanfare because it's the tactical portion.

The U.S. Air Force houses all of the MCEs within the United States, spread across various bases, with the majority of squadrons at Creech Air Force Base in Nevada and Cannon Air Force Base in New Mexico. Additionally, there are Air National Guard units spread across the U.S. that fly the MCE portion of a flight. The British Royal Air Force has adopted a similar method of remote split operations, with a squadron co-located with USAF RPA squadrons at Creech (39 Squadron) and one located at RAF Waddington in Lincolnshire (13 Squadron).

This will likely continue for the foreseeable future since there are several advantages to keeping a small footprint in the dangerous war zone while having the majority of your personnel work from the safety of the home country. One major advantage of RSO is that it keeps all the pilots, sensor operators, and intelligence analysts out of harm's way. If you know anything about fighting a war, then you know that General Patton's speech prior to the U.S. Army leaving Africa during World War II hits the heart of the matter: "No dumb bastard ever won a war by going out and dying for his country. He won it by making some other dumb bastard die for his country." That is the essence of the benefit of remote split operations: Kill the enemy and preserve your own forces.

Another major advantage to the RSO model is the cost savings. By keeping all these warriors at home remotely flying combat missions, the Air Force saves money in mobilization, housing, feeding, force protection, and the cost of additional combat-related pay. If your personnel can support combat operations from home just as effectively, why send them forward where it costs the institution more and they run the risk of exposure to danger? A tangentially related savings is boots-on-the-ground numbers (referred to as BOG in the military). Military operations throughout the world are often politically limited to a certain number of troops deployed in foreign territory, and those numbers are highly scrutinized to ensure efficiency and efficacy. Remote warriors don't count against BOG, but contribute significantly to the war effort, which for a commander trying to decide which troops to let into the fight is both a significant benefit and a substantial relief.

At first glance RSO sounds like the perfect employment model; commute to work, fly a combat mission, kill some bad guys, commute home in time to catch your daughter's soccer game. Unfortunately, every innovation has its unanticipated and unintended consequences, and RSO is no exception. The cost in this case, however, is paid by the employee's mental health as opposed to the employer's dollar. The cost is paid in operational fatigue, burnout, cynicism, stress, retention of quality personnel, family troubles, sleep disruption, shift work, commuting to combat, the daily transition to and from a combat mentality, negative disruptive emotional responses to killing, and this perpetual "deployed in garrison" feeling. I will cover all these challenges in greater detail, but it's important to note that they have been recognized by the Air Force through various studies and marginal improvements have been made. The employment method of RSO is a major contributor to the challenges RPA crews experience.

Transitions: Commuting to Combat

The problem with RSO resides in what I call transitions. It is both a physical and mental process. When traditional warriors deploy to a foreign country for combat, they must physically and mentally transition twice. During the period leading up to a deployment, warriors are already starting to detach from those around them mentally, preparing the mind for what it is about to go through. Oftentimes they write a will or a final letter to be opened by loved ones in the case of their demise in combat; effectively, mentally cutting the cord with those they are leaving behind. This process could not be more natural, and it has been practiced throughout the history of human warfare; mentally detaching is a survival mechanism that helps the mind focus entirely on the task at hand once deployed: fighting and surviving.

During the deployment, the warrior is surrounded by similar individuals from their unit and they experience and survive hardships together throughout the duration of the deployment, forming an intense bond with one another. As the end of the deployment draws close, the warrior's mind starts to wander to thoughts of home; what he will eat when he returns, who he will see, what he will do. This usually happens within the last month of a deployment and it is extremely dangerous. During this period, it is critical that a commander reinforces and emphasizes the mission, safety, and fighting complacency in an attempt to keep the warrior's mind focused. As most commanders can testify, there is nothing more real and demanding during the deployment than the struggle to keep the warrior's mind in the present, in the fight, and not distracted by thoughts of home.

After the deployment ends, the unit goes through a warrior transition period. How this transition occurs differs from unit to unit, but it's a transition period from the high-stress, high-operational tempo of combat to life back at home. It is a critical incident debrief. It is warriors sitting around talking about their experiences with one another and oftentimes with other professionals such as chaplains, doctors, or psychologists. It's cathartic and healing and exactly what the warrior needs for the sake of their mental health before returning home. It's similar to the gathering around the campfire at the end of a day's battle that warriors have done for thousands of years. Two transitions for every deployment.

Now imagine rather than two transitions per deployment, two transitions per day. Drive to work, get ready for combat, fly a combat mission, mentally prepare to transition back to family life, and then drive home. Wake up the next day, wash, rinse, repeat—for four straight years!

The transition period becomes compressed to a short five- to thirty-minute commute home. The ability to compartmentalize between combat and life becomes strained. Routine things that happen in life become trivial compared to the fact that you just killed a guy in front of his family with a Hellfire missile four hours ago. Your mind simply can't transition that quickly, that frequently, or that drastically. Sleep is its only respite to heal. But sleep is often disrupted by family or the nature of doing shift work. At first this stress may be minimal, but eventually it can accumulate and over time it can turn into chronic stress; wearing down your mental armor, your resiliency, and leaving you more vulnerable and susceptible for trauma to seep in and take hold. A seasoned Reaper sensor operator named Matthew described the complexity of transitions, of stressors in life and combat coexisting in a single day:

> On my first strike, we were following an individual and two of his cohorts in Afghanistan, a fleeting target of opportunity of three armed enemy fighters. Our JTAC was very intense and very spun up. He got louder and more excited the further along we went into the engagement process and it was only adding to our stress level. As our targets walked through a village they were walking in and out of green terrain [green terrain is an area free of collateral damage where it is ideal to strike the target]. My pilot muted the radio to the JTAC and said, "Our JTAC is amped up, listen to my voice and my commands only."

Then, when the targets moved into green terrain, the JTAC started yelling, and he cleared us hot six times in a row to engage the targets. Of the three individuals we were targeting, I aimed for the middle of the pack between where they were walking. The missile was launched but my boresight [correlation between where the sensor is looking and where the laser is pointing] was apparently off. Instead of the missile landing directly between the three targets where I aimed, it landed eleven meters in front of them and sprayed all three of them with fragmentation from the missile. None of them died instantaneously. Instead, I watched them fall to the ground and roll around in intense pain from the fragmentation.

Almost immediately, twenty locals from the village began surrounding these three individuals and started providing first aid. The locals then made makeshift stretchers and we watched them carry the wounded men for a mile down the road to a hospital. The whole way, I'm basically watching three people bleed out on stretchers because of an error. We continued to watch the hospital until around three a.m. when we saw three corpses coming out of the building and then the JTAC declared the mission over.

Afterward, we did a debrief with the crew and then with the JTAC and commanding officer and then with the LRE. It turns out that the LRE crew never boresighted the laser before they handed us the airplane, which caused a mismatch between where the missile went and where I aimed. As a result of this mission, procedures were changed across the entire Air Force enterprise to do a post-mission boresight of the laser to ensure that it was done before the next mission.

By eight a.m., I arrived at home at the end of a very long day, quite possibly one of the worst days in my life, and my wife handed me my one-month-old colicky son. I just ended three human lives for the first time in the most vicious way possible with lots of human suffering, had a five-minute commute home to shut that off in my mind, and then dealt with a colicky baby.

This isn't a unique anecdote. Almost every pilot, sensor operator, or intelligence analyst I interviewed shared a story of two worlds colliding, with them in the middle sorting it out. One British Reaper pilot told of how he had to be removed from flight duty for a month because he had been averaging two hours of sleep a night over a period of three months. His mind, as he put it, was mush.

The problem of frequent transitions is not a secret. Much has been written about this odd manner of fighting a war. Dan Gettinger, in an article titled "Burdens of War: PTSD and RPA Crews," stated:

> In addition to the possible reasons cited by previous reports, the newest study suggested that RPA crews face challenges that are unique to RPAs and "telewarfare." One such source of stress is the "social isolation during work, which could diminish unit cohesion and thereby increase susceptibility to PTSD." Whereas service members in combat theaters may depend on their unit for support, the individualized nature of piloting RPAs, combined with the stresses of shift work and the daily transition between war and peace, creates unique mental and emotional pressures for RPA crews.

A National Public Radio segment titled "Report: High Levels of Burnout in U.S. RPA Pilots" that first aired on the show *Morning Edition* described these challenges as early as 2011:

> The Air Force cites several reasons for the elevated stress levels among RPA pilots. First is the dual nature of this work: flying combat operations or running surveillance in a war zone, and then, after a shift, driving a few miles home in places like Nevada or New Mexico, where a whole different set of stressors await. The Air Force says switching back and forth between such different realities presents unique psychological challenges.

In 2013, Dr. Jean L. Otto and Air Force Captain Bryant J. Webber from the Armed Forces Health Surveillance Center and Henry M. Jackson Foundation for Advancement of Military Medicine published the results of a study in an article titled "Mental Health Diagnoses and

Counseling Among Pilots of Remotely Piloted Aircraft in the United States Air Force." The study, which looked at the mental health diagnoses of 5,256 manned aircraft pilots and 709 RPA pilots during the period of October 1, 2003 to December 31, 2011, determined there was no significant difference between the rates of mental health diagnoses, including post-traumatic stress disorder, depressive disorders, and anxiety disorders between RPA pilots and traditional manned aircraft pilots. Otto and Webber went on to describe some of the challenges confronted by RPA pilots:

> In particular, RPA pilots have to deal with a lack of deployment rhythm and of combat compartmentalization (i.e., a clear demarcation between combat and personal/family life); fatigue and sleep disturbances related to shift work; austere geographic locations of military installations support-ing RPA missions; social isolation during work, which can diminish unit cohesion and thereby increase susceptibility to PTSD; and sedentary behavior with prolonged screen time.

One of the instrumental advocates for RPAs in the Air Force has been Brigadier General James "Cliffy" Cluff. In June 2018 he assumed the role of director of remotely piloted aircraft and Big Wing ISR aircraft for the Air Force. Operationally, he has flown it all: the Predator, Reaper, Global Hawk, and the RQ-170 Sentinel. From May 2013 to June 2015 he commanded the 432nd Wing and 432nd Air Expeditionary Wing at Creech Air Force Base, the first Wing to operate RPAs exclusively. While serving as the 432nd Wing Commander, Cluff was quoted in a *New York Times* article titled "As Stress Drives Off RPA Operators, Air Force Must Cut Flights," referring to the challenges associated with transitions:

> Having our folks make that mental shift every day, driving into the gate and thinking, "All right, I've got my war face on, and I'm going to the fight," and then driving out of the gate and stopping at Walmart to pick up a carton of milk or going to the soccer game on the way home— and the fact that you can't talk about most of what you do at home—all

those stressors together are what is putting pressure on the family, putting pressure on the airman.

This must be brought into perspective by noting that troops in any war in history would gladly have traded places! In fact, in recent years, Marine RPA pilots have actually been transferring to the Air Force to fly Reapers at an alarming rate. The reason? Killing combatants by day and being home in time for dinner sounds like a pretty appealing warrior lifestyle when you are accustomed to harsher conditions.

These rapid transitions between home life and combat are not without precedent in the military or law enforcement agencies. A similar stress was experienced by bomber crews in England in World War II (especially British crews), but modern RPA crews do not confront the personal, physical danger those World War II bombers faced. None of those bomber crews would have volunteered to move to the forward edge of the battlefield, in order to avoid these complications, when they could fight the war from home. During the Kosovo War, Air Force B-2 bombers flew thirty-hour combat missions bombing Serbian air defense assets, starting and finishing each mission from their base in Missouri. And police face the same conundrum: on the "mean streets" every day in life-and-death circumstances, and then home for dinner; yet most police officers have no great difficulty dealing with this dilemma for their entire career. Nevertheless, it is a valid concern and an irrefutable stressor in the lives of these RPA warriors.

The Ultimate Oxymoron: Deployed in Garrison

There are many ways to describe this strange occurrence of commuting to combat, but the description most prevalent in the Air Force is "deployed in garrison." Although garrison technically means a force of troops in a fort to defend it, in military parlance it means "at your home base," or "at home"—the opposite of being deployed or in the field. "Deployed at home" is obviously an oxymoron, but it accurately describes the environment RPA crews are fighting in every day. In fact, some of the RPA bases in the United States have signs posted in the Reaper Operations Center that say "You are now entering CENTCOM." The Central Command—or CENTCOM—area of operations includes the majority of countries where the Air Force is flying combat air patrols such as Syria, Iraq, Yemen, and Afghanistan. This sign is not ostentation, it's deliberate. The statement is a reminder to assume a combat mindset, and although you will remain physically distant from the fight your cognitive distance is about to get really close and intimate. Air Force RPA pilot Dave Blair, and Karen House, a licensed professional counselor, who served with the military for ten years, describe their take on transitions and being deployed in garrison in their phenomenal article "Avengers in Wrath: Moral Agency and Trauma Prevention for Remote Warriors":

Reaper flight crews transition from a combat mindset to a domestic one in the space between leaving the cockpit and arriving home. Pilot and sensor operators are currently termed "Deployed in Garrison," a term for RPA Airmen who, on a daily basis, virtually enter hostile territory where others' lives are on the line; most operations centers have a sign that says *The time between killing and coaching a child's soccer team or having breakfast with a spouse and children could be less than an hour.* This ability to switch between combat and domestic mindset is an untested appliance; we have not had enough time or experience to calibrate our baselines to predict when the capability of such a mental switch will expire from over-utilization. We do know that rest will prolong the crews' ability to flip that switch—perhaps (hopefully) indefinitely.

In the article "The Dimensions of Contemporary War and Violence: How to Reclaim Humanity from a Continuing Revolution in the Technology of Killing," the psychiatrist Robert Jay Lifton discussed how being deployed in garrison affects RPA crews. He described a typical day for an RPA crew as long hours of remote warfare followed by a quick drive home, with two separate and distinct environments that the mind is exposed to in a single day. Lifton theorized that what enabled this daily transition between garrison and combat was an individual's ability to develop a functional second "self" to adapt to the different environments. Essentially to develop two identities: a normal "self" at home and a second "self" that is at war.

If this is indeed possible, one can imagine the long-term effects on the psyche of an internal daily struggle between two "selves." Although no one I interviewed used the same terminology of creating a second "self" to adapt to the constant transition between combat and garrison, almost everyone referred to switching on a warrior mindset or combat mentality at the beginning of their shift. Perhaps this hypervigilant warrior mindset could be viewed as a second "self."

It is safe to say that there is no frame of reference for those who have experienced war in the same manner RPA warriors have over the last several decades. There is no equivalent job that commutes to work and kills people on the other side of the planet. Law enforcement officers are the closest equivalent to RPA warriors; however, there are two main differences between them.

First, the violence experienced by our police officers and border patrol agents in the line of duty is up close and personal. Second, on average our RPA warriors experience violence more frequently than law enforcement officers. Of the 254 RPA personnel I surveyed, 138 of them (54 percent) had been involved in eleven or more strikes in their career, forty-four (17 percent) of them had been involved in fifty or more strikes. While it isn't a given that a police officer will be involved in a deadly confrontation in their career, 96 percent of all RPA warriors I surveyed had been involved in a least one deadly encounter—a statistic that is close to near certainty. A similar Air Force study led by the clinical psychologist Dr. Wayne Chappelle revealed that of the seventy-one Air Force RPA aircrew interviewed, forty-five (61 percent of them) had participated in seven or more strikes.

It is important to note here that "strike" is not necessarily a euphemism for "kill." As mentioned before, another historical parallel to modern RPA crews would be World War II British bombers flying from England to attack targets in Germany. These bomber crews knew that they were probably killing a lot of people with each bombing run, many of them innocent civilians killed as collateral damage. Additionally, both bombing runs and RPA strikes are sometimes directed at objects (buildings, bunkers, etc.) as opposed to individuals. In both instances to call these "kills" vs "strikes" or "bombing runs" would be imprecise.

The difference between bomber kills and RPA kills is, first, the precision of RPA strikes, which almost completely eliminate collateral damage; this can be seen as a positive from a mental health perspective. The second difference is the intimate nature of RPA kills, which can be seen as a negative.

This intimacy is the reason I suggest that RPA strikes could usefully and practically be divided into different parts. In infantry combat, the soldier who kills the enemy should never be the one who searches the body. In the same way, the RPA strike could be broken up into smaller, more emotionally digestible pieces. One crew can do the prep work, another makes the strike, and another does the BDA or aftermath assessment. This diffusion of responsibility in the killing process should be considered now, or it might be the first step to be taken if the psychological toll on RPA crews needs to be addressed in the future.

Coming back to the parallels between law enforcement and RPA operations, there are, fortunately, warriors who can assist with this transition comparison since they have served both as police officers and as RPA warriors. Tim Stack was initially a Predator sensor operator starting in 2007 and then transitioned to be a Reaper sensor operator in 2012. He performed as a weapons and tactics instructor for a California Air National Guard squadron for five years. He was also a full-time police officer for the Los Angeles Police Department for more than twenty years before joining a police force in Texas. He is one of those warriors with a unique perspective on transitions and both the similarities and differences between commuting to combat as a police officer and as an RPA warrior.

There is a similarity. In the Air Force, from the parking lot to changing into your flight suit, to getting into the operations center, a switch is turned. It becomes pretty real, pretty fast. As soon as you are off work, you leave and go home, and the transition is longer and more drawn out over the whole drive home. Especially if you struck that night. You are reliving it the entire way home and when you get home your family asks, "How was your day?"

Police work is kind of the same. There is a short transition to get to work and into the role of an officer. There is a routine of it. You know you are going to make traffic stops today, you know you are going to answer radio calls today, the thing you don't know is if it is going to be violent today. This month alone I've stopped 111 cars. Each time there is an opportunity for it to go sideways. I change into my uniform at the station and before every shift I look myself in the mirror and say, "It's time to go to work." It's a mindset. There are a lot of similarities to it, as far as changing mindsets to go to work and then having that drag at the end of the shift to change back to civilian life.

The long-term effect of this daily transition to and from combat on the mental health of remote warriors is yet to be seen, but the daily transition is not the only unintended consequence of remote split operations. A number of additional unintended consequences have materialized, including a lack of time for adequate debriefs, isolation, stresses of shift work, inability to talk about struggles with family or a support network, and being deployed in garrison indefinitely with no dwell time or respite from operations.

Shift Work: The Worst Thing We Could Do to a Warrior

Air Force RPA personnel operate in shifts, split between a day shift and a mid-shift. Crews rotate between these shifts on a routine basis. The frequency of the rotations between shifts varies by unit, but typically rotations occur every six weeks. The day shift in the U.S. coincides with nighttime hours in the CENTCOM area, a period with relatively less enemy activity. The mid-shift coincides with daytime hours overseas and is usually the shift most likely to have a strike.

So not only is your circadian rhythm, the body's natural sleep cycle that aligns with daylight hours, 180 degrees out during the mid-shift, but you are also more likely to participate in a critical incident. Much has been written on the challenges of shift work and its effects on the human body. I'd like to tell you that routinely altering your circadian rhythm has no effect on a person, but unfortunately that couldn't be further from the truth.

Frank Scheer, PhD, a neuroscientist at Harvard Medical School and Brigham and Women's Hospital in Boston, states, "There is strong evidence that shift work is related to a number of serious health conditions, like cardiovascular disease, diabetes, and obesity. These differences we're seeing can't just be explained by lifestyle or socioeconomic status." The term *shift work* generally refers to rotating shifts, and rotating shifts have been connected with numerous problems such as stomach issues, ulcers, depression, and an increased risk of accidents or injury. Research reveals that rotating shifts is harmful to families: families can handle day shift or night shift, but over the long haul most families cannot handle rotating shifts. Shift work in the RPA community was born out of the insatiable demand for ISR and, unfortunately, there is no end in sight to this challenge.

It should be noted that here, again, a law enforcement parallel might be useful. Faced with the need for twenty-four-hour coverage, for more than a century, law enforcement agencies worldwide have experimented with every possible aspect of shift work. The answer in that community keeps coming back to a simple model: Bid for your shift based on seniority.

If you want to stay on night shifts, good for you. It takes up to a full year for the human body to fully adapt to the night shift, and every change in shift creates a period of "jet lag" that significantly reduces performance and harms families. Long periods of rotating shifts have a health impact that can literally take years off your life. If you want to change to another shift when the opportunity arises, by all means do. You have earned it. You have the seniority. This is truly a fair system that can reduce the physical and psychological toll of shift work.

A Killer on a Lonely Island

RPA crew members can experience isolation as a result of being deployed in garrison and switching between day shift and mid-shift. One young Air Force lieutenant and Reaper pilot told us he didn't even feel like he was in the Air Force due to the strange, isolated environment he worked in. He commuted to and worked on a base with only other RPA crew members. He rarely saw anyone outside of his squadron and even less frequently saw other Air Force personnel who weren't part of the RPA community. He was disheartened by it and had little reason to feel connected to the service.

Even within the squadron it can be extremely isolating. Due to the nature of shift work and supporting operations 24/7, an RPA squadron is never able to assemble the entire unit at one time. It is typical for pilots or sensor operators new to a unit not to meet all of the other aircrew in the squadron within their first six months. We know that peer support is the glue that binds a unit together and that develops as a result of its members actually knowing each other, sharing hardships in challenging environments, and hanging out together outside of work. So this lack of peer support is unfortunate.

Air Force Reaper pilot Major Johnny Duray eloquently described this environment of isolation in the article "Forever Deployed: Why 'Combat-To-Dwell' Reform for MQ-9 Crews Is Beyond Overdue":

> In a best-case scenario, an airman breaks from flying combat sorties every four to five days. Flying time plus additional duty and administrative work commonly translate into 12-hour days and sometimes more. Fewer breaks and longer days are frequent. . . .
>
> . . . once every four and a half months, an airman is off on a Friday and Saturday evening with the chance to socialize and form friendships in a manner consistent with most airmen in their early twenties.

Want to Talk About It? Are You Good?

Debriefs occur after every flight. If a strike occurs during a flight, the crew can expect the debrief to be longer and to have additional administrative work to complete. That's if everything goes well and according to plan. If everything doesn't go according to plan, the crew's day just got even longer than the already scheduled twelve-hour shift. The post-mission debriefs are professional but tend to focus entirely on the mechanical aspects of the strike, covering topics such as the communication with the JTAC, the timing of the strike, evaluating whether the target was struck in the most lethal and effective manner, and what could be improved on for next time. Rarely is the emotional side of the mission—defined here as an individual's psychological and physiological response to taking another human life—discussed during a debrief.

This is partially a culture thing. The military develops a culture for killing, but not a culture for dealing with the act of killing. Oftentimes that burden is left to the individual to figure out by themselves. There are mental health professionals, counselors, doctors, chaplains, and peers who are available to assist, if desired, and the military has made great headway in this area in the last two decades of fighting wars, but the system remains flawed. The system relies primarily on self-identifying and self-reporting, which runs counter to the culture of toughness.

Many RPA warriors told me that the help offered is merely lip service. It consists of a psychologist stopping by after a strike and asking the entire crew at the same time, "Are you good? Is everybody okay?" Even if a warrior isn't okay, they're unlikely to admit it to an outsider in front of the crew they just fought with. The warrior does not see the benefit at that moment of debriefing with a psychologist. It adds one more thing that must be accomplished at the end of a very long and stressful day and is one more roadblock before they can go home to be with their family or just get away from work. Remember, there are two worlds competing for this warrior's time in any given day: the combat world and the home world.

So the debrief suffers as a result of these transitions. Seventy-one percent (181 of 254) of the people I surveyed said that it wasn't common for a discussion with a mental health professional to take place post-strike. It was most common in the Air Force (45 percent of the time) and almost never happened in the Army and Marine Corps (5 percent of the time). The Air Force has made extensive efforts to improve in this realm and I expect that number might be much higher now. However, no such initiative exists in the other two branches of service. And we know the extreme importance of the critical incident debrief for the mental health of someone who has gone through something as stressful as taking a human life.

The First Rule of Fight Club

Compounding the sense of isolation is the secretive nature of the RPA crew's work. Due to the classification level of missions, crews are not allowed to discuss events that occur during the mission with people who don't have the appropriate security clearance and the need to know the details. A crew member could have an extremely emotionally challenging day, particularly if he is struggling with the circumstances of a strike, and not be able to share it with their spouse or support network outside of work. Some crew members may also not want to burden their loved ones with the details of gruesome, vivid incidents such as beheadings, civilian casualties, or witnessing a person blown to pieces by a Hellfire missile.

Some do choose to share with their spouse the news that they had a strike, obviously omitting classified details of the mission. In those instances, everyone who told us that this was part of their routine found their spouse to be supportive and helpful in the process of dealing with the situation. In all likelihood, though, due to the increased demands of time spent debriefing and completing administrative work after a strike, a spouse already knows before their RPA warrior comes home that something happened. Fortunately, as previously touched on, the Air Force recognized this gap and took measures to correct it by embedding teams of mental health professionals with appropriate security clearances in operational squadrons, and that gives some hope that this will be the new standard across the enterprise.

The Never-Ending Deployment

All of these problems are related to the method of RSO, and as described by both personal anecdotes and medical professionals, this employment method can have direct, negative impacts on the performance, health, and overall well-being of the participating warriors. There is, however, one more significant overarching problem: no respite from operations. A traditional military unit experiences a deployment rotation cycle. Prior to a deployment, a unit endures a preceding period of training to prepare it for the challenges of combat, then there is the deployment, and after the deployment there is a period of respite, a time where the members of the unit unwind, relax, and recover from the rigors of deployment or combat.

The time between deployments is referred to as dwell time and is considered healthy and necessary for the long-term sustainability of the deployment cycles that have been necessitated by fighting the longest wars in the history of the United States. The desired deploy-to-dwell ratios within the service are 1:2 or 1:3. As an example, a 1:2 deploy-to-dwell ratio would give a unit six months deployed and twelve months at home. A 1:3 ratio would see a unit deployed six months out of every two years. A higher number on the right obviously means more time home. If the number on the left of the ratio is higher than the number on the right, that means the unit is deployed more than it is home.

For RPA units that deploy forward to support operations, it has been a challenge to maintain a 1:3 or even 1:2 deploy-to-dwell ratio due to the huge appetite for RPA support and a shortfall of units able to provide it. The solution was often to reduce the deploy-to-dwell down to an undesirable ratio. This led to burnout, operational fatigue, and retention issues. As an example, from 2003 to 2008, the Marine Corps' only two

RPA squadrons (VMU-1 and VMU-2) were both on a 1.4:1 deploy-to-dwell ratio (seven months deployed, five months home) for five years straight in Iraq, with one month of overlap in-country for turnover. The five months at home were spent preparing for the next deployment, with very little time off. Those two Marine Corps RPA squadrons were the most deployed units in the Department of Defense for five years running. The full impact of imposing this schedule on a community took years to fully realize, but the impact was reflected most immediately in the failure to retain talented Marines who were burned out and left when their contract was over. Unfortunately, the situation was not unique to the Marine Corps, as the Army experienced similar challenges with its RPA units.

But what about units that don't "deploy"? Where does the Air Force fit into this model conducting remote split operations? The short answer is worse off than deployable units when it comes to respite. There are two schools of thought about this issue. Some would argue that, due to the employment model of conducting operations remotely from the United States, if RSO were expressed in a deploy-to-dwell ratio, the number on the left would be zero and the number on the right would be based on the number of years that unit had supported operations, something like 0:15. Conversely, if you accept that what our Air Force warriors are engaged in on a daily basis should be considered combat operations, then RSO is a constant state of deployment without a dwell period and the ratio becomes 15:0. Air Force Major Johnny Duray captured the heart of the problem well in "Forever Deployed":

> At the same time, the MQ-9's unique combat capabilities drove demand for this platform through the roof. This demand, along with fundamental misunderstanding of RPA combat operations shift-work, cultivated the false notion that MQ-9 operators can conduct combat operations indefinitely without the need for operational respite. . . .
>
> But, it's argued, this is combat. No one takes days off while deployed. This is true, but no other community conducts combat operations for five, six, or even seven straight years. Other military units sprint, rest,

then sprint again. No one keeps sprinting for years at a time—except the Air Force's remotely piloted aircraft community.

The reality is that the RSO employment model doesn't fit neatly into a comparison with deployable units, and we get bogged down debating semantics over whether someone is "deployed." What we should be considering is whether the mind needs a break from high-stress, high-tempo operations that involve killing enemy combatants, regardless of where the unit executes it from. There is, again, a parallel to law enforcement officers, who often continue their "mission" for virtually every working day of their career. Even when off duty, most police carry their weapons and are prepared to take action, if needed, to save lives 24/7. If the law enforcement community can do this for an entire career, can RPA crews do the same?

Looking at RPA "deployment issues" and attrition rates, there is cause to believe that something needs to change. If the employment method of conducting combat operations from the United States can't or won't be changed, then there at least needs to be a break in the deployment cycle, a rest or dwell period added into it.

Hub and Spoke: The Wagon Wheel of UAS

When we discuss how RPAs are employed, one major distinguisher is whether they are controlled within line of sight (LOS) or beyond line of sight (BLOS). RPA units outside the Air Force must deploy to the area where the RPA is flown within line of sight of the aircraft. For commercial RPAs, "line of sight" actually means remaining within visual sight of the pilot. For military RPAs, however, line of sight refers to the distance the aircraft can receive and transmit its control signal to and from a ground radio while airborne. This distance extends well beyond the visual range. Beyond line of sight implies that the RPA can travel beyond the range of the radios on the ground used to control the airplane and that the control link is transferred to another method, usually satellite communications.

Yet another creative military innovation born of necessity was what is referred to as a "hub and spoke" employment method. The "hub" is similar in nature to an LRE site in the RSO model: a place where the aircraft are prepared, launched, and handed off to a "spoke." The spoke is another ground control station and aircrew that receives the aircraft through a control station transfer, flies the tactical portion of the mission, and then hands the aircraft back to the hub.

This concept initiated from conducting static, counterinsurgency operations in Iraq. The benefit to "hub and spoke" is that you can consolidate all your logistics, maintenance, force protection, and intelligence support at the hub, and each outlying spoke consists of just a ground control station with the aircrew, normally embedded directly into the operations center of the unit they are working with. The other major advantage is that you can extend the range of the aircraft. As an example, if the line-of-sight range of an RPA goes out to fifty miles, you could place another ground control station, a spoke, one hundred miles away from the hub and fly another fifty miles beyond that. You have effectively tripled your range and doubled the area you can cover by positioning a spoke at maximum line-of-sight range. This technique was used extensively in Iraq and Afghanistan.

We routinely use robots to remove people from harmful situations, so it's counterintuitive to think that in some instances we do the opposite and actually put people in harm's way so they can employ a robot. In March 2017 Marine Captain Joshua Brooks was deployed to a country that was fighting the Islamic State and other violent extremist groups. The limited range of the RPA his unit flew from the hub didn't allow it to cover the entire area where the enemy fighters were hiding.

In order to increase the range of the RPA and gather intelligence and target the Islamic State fighters, Captain Brooks and two other Marines embedded with a special operations team and established a spoke site at a small outpost in the heart of enemy territory, protected only by a foreign country's army and the limited weapons they carried with them. Over the course of the next six months, the intelligence that this three-man spoke team delivered to the host nation's army enabled multiple artillery strikes on the enemy and denied a sanctuary that the enemy fighters had come to rely on. Something as simple as a spoke team and the information they provided changed the dynamic of the battle. As ground control stations continue to miniaturize and become more portable, this trend of embedding the RPA pilot on the front lines with the people they are supporting will continue to increase.

Since units that employ line-of-sight RPAs must deploy to the country they are fighting in, they do not have the same issues as the Air Force regarding transitions, isolation, an absence of respite, sleep disruption, and shift work. The trade-off, however, is more personnel in harm's way, the financial cost of deploying and sustaining troops overseas, and the added stress of spending time away from family and home.

Albeit the above costs and consequences of RSO can be negative in nature, the method of employment does not cause negative emotional responses to killing with RPAs directly. Instead, the RSO method can contribute indirectly, as we have seen, by wearing down an individual warrior's mental armor, making them more susceptible to trauma. The factors outlined in this chapter are critical aspects that require further research and attention in order to sustain RPA operations across the decades to come. In the next chapter I discuss how we kill with RPAs.

A Spectrum of Responses to Killing with RPAs

Chapter One

How Do We Kill with RPAs?

1. A robot may not injure a human being or, through inaction, allow a human being to come to harm.
2. A robot must obey orders given it by human beings except where such orders would conflict with the First Law.
3. A robot must protect its own existence as long as such protection does not conflict with the First or Second Law.
> —Isaac Asimov's Three Laws of Robotics, *I, Robot*

Tragedy in Dallas: The First Robot-Assisted Kill in the U.S.

On July 7, 2016, Micah Xavier Johnson, a former Army reservist who served one tour in Afghanistan as a carpenter and did not see any combat, ambushed and murdered five Dallas police officers and wounded nine others. The officers had been assigned to keep the streets safe during a rally protesting the killings of Alton Sterling in Baton Rouge, Louisiana, and Philando Castile in Falcon Heights, Minnesota. Sterling and Castile were both killed in officer-involved shootings the weeks prior that led to protests. That fateful day, Johnson set out with a rifle and pistol intent on murdering as many police officers as possible. The ambush started at 8:58 p.m., at the conclusion of the rally, with the killing of three officers and the wounding of three other officers and a civilian, when Johnson opened fire from the street. Johnson then murdered another officer, shooting him multiple times from behind before breaking into a nearby El Centro College building.

From a second-story window of the college, Johnson shot and killed another police officer across the street at a 7-Eleven. He then barricaded himself in a room where he and Dallas SWAT officers exchanged more than two hundred shots during a standoff. Negotiations for his surrender continued for hours but broke down when Johnson claimed he had planted bombs throughout the city, began laughing and singing maniacally, and stated that he wanted to kill more officers. By 2:30 a.m. Dallas police chief David Brown was out of options that wouldn't lead to the death of more police officers and he authorized the use of a bomb disposal robot armed with one pound of C-4 explosives to detonate against a wall next to where Johnson had barricaded himself.

The plan worked, killing Johnson and ending the greatest loss of police officers' lives in a single attack since September 11, 2001. It was the first time in American history that a robot was used by law enforcement to kill an active mass murderer. Reaction to the Dallas Police Force's actions varied in the media, with some hailing Chief Brown's decision as necessary and pragmatic while others condemned the decision, claiming that it denied Johnson the right to due process. In January 2018, Dallas County district attorney Faith Johnson (no relation to the killer) released a statement noting, "As with all officer-involved cases, this case was presented to the Grand Jury to determine if the use of force by police was reasonable given the unique facts and circumstances of the case presented." The grand jury ultimately declined to bring charges against the police officers who had used the explosive-laden robot to bring an end to the situation, finding their actions prudent and reasonable.

The tragedy in Dallas brought to light many of the challenges associated with killing with the aid of a robot. If any one of those responding police officers in Dallas would have been the one to kill this mass murderer with their service pistol or rifle, they would have rightfully been hailed a hero. In fact, society demands it of our nation's protectors to immediately respond to a wolf that threatens the flock. But when the Dallas police killed the shooter with the assistance of a robot, somehow this was viewed as a different and questionable use of force. Two points are worth noting. First, the decision to kill the shooter with a robot was made by a human with the appropriate authority to do so. Secondly, the robot was not a lethal autonomous robot; it wasn't the Terminator 2000 unleashed on the streets of Dallas. It was just another tool altered for the use of humans to resolve a deadly situation, which is exactly what society demands of our law enforcement warriors.

Just Another Tool: Not the Terminator

Some members of our society harbor a mistrust of using a robot—whether it be the Terminator 2000 or R2-D2—to aid in resolving a deadly encounter. There may be a misconception that robots, and therefore RPAs, are autonomous entities killing on the battlefield. But they are not; and, fortunately, we have not reached that extremis from a technology, legal, or policy standpoint. Rather, RPAs and robots are just tools executing programmed algorithms, controlled by the highly trained professionals piloting them.

We must remember that Isaac Asimov's Three Rules of Robotics are based on the writings of a science fiction author referencing an independent-acting, autonomous robot, not the robots that exist today, which are just an extension of the humans employing them. A key difference between Asimov's conception of a robot and today's robots is that the accountability and decision to kill using a robot resides with a human involved in the decision-making loop. When we discuss how we kill with RPAs in the military, this process is no different. The RPA is just a tool, remotely employed by a human, to augment combat.

Killing with an RPA can be done either directly or indirectly. By directly I mean employing a weapon off the rail or bomb rack of the RPA, whereas indirectly is using the RPA to cue another asset to kill an enemy combatant; e.g., using the RPA's laser or sensor. One may argue that killing indirectly isn't actually killing. On the surface, that is technically true, but what I have found through my research is that even crews who kill indirectly using an RPA may still be affected by the experience. A crew piloting an unarmed RPA still feels responsible and accountable for the outcome of their mission. And that is actually a good thing.

This level of accountability is a reflection of a crew's professionalism, training, and involvement in the mission to ensure that only correct, accurate, and legitimate military targets are struck while keeping collateral damage mitigated or avoided regardless of where the actual ordnance is delivered from. This ownership of killing without actually pulling the trigger should come as no surprise. *On Killing* discussed this feeling of ownership experienced by a Vietnam veteran named Dave who was not a trigger puller but felt the burden anyhow:

> There is no definitive distinction between the guy pulling the trigger, and the guy who supported him in Vietnam. They may not have killed, but they were there in the midst of the killing, and they were confronted with the results of their contributions to the war.

The same is true for RPA crews. They are in the midst of the killing and they are confronted with the results of their contributions to the war; it doesn't matter that their actions are mediated through a high-fidelity, zoomed-in camera.

Degrees of Participation in the Kill: Shooters, Assisters, and Watchers

How we use an RPA to kill can be divided into three categories based on a crew's degree of participation and involvement in a strike. Let's use the basketball analogy of shooters, assisters, and watchers. Shooters take the shot, assisters pass the shot off to someone else, and watchers are those players on the court who don't touch the ball but watch the shot go down. Shooters are those RPAs that are large enough to be armed and used to kill directly from their own platform, such as Predators, Gray Eagles, and Reapers. Assisters are RPAs that carry an onboard laser target designator used to mark the shot for another platform or to adjust indirect fire missions.

Watchers are those RPAs that find a target, call someone else to shoot it, and then report the aftermath of the strike, but do not mark a target with a laser or employ a weapon against it. Watchers can also be non-aircrew members who assist during the shift, such as intelligence analysts who work outside of the GCS. Irrespective of role, shooters, assisters, and watchers can all have the same level of exposure to the destruction of the target since they are all involved in the battle damage assessment. However, each role has a different level of responsibility and accountability in the actual shot/kinetic engagement process.

Given this dispersion of responsibility and accountability with the various roles, crew positions, and degrees of participation, it should follow that the response to killing could vary among crew members. To explore this concept, I surveyed 254 people across all the U.S. military services who collectively had employed all types, sizes, and capabilities of RPAs. Two-hundred and forty-three of my survey participants had combat experience using an RPA to kill either directly, indirectly, or both, as is often the case with an armed RPA aiding another platform. I also discussed this question during many of the interviews I conducted. Additionally, I looked at different positions within a crew—pilot, sensor operator, and intelligence analyst—to determine if there were any differences in how they responded to killing. I will discuss the entire spectrum of responses in the next chapter, but will focus for now on the responses when killing directly or indirectly, as depicted in Tables 1 and 2.

Table 1
Engagement Method and Self-Diagnosed Trauma

Engagement method	n	% of over-all respondents	% of Self-diagnosed trauma "yes"	% of Self-diagnosed trauma "maybe"
None	11	4.00%	9.09%	0.00%
Directly (shooters)	107	42.00%	14.02%	28.04%
Both (shooters)	79	31.00%	12.66%	17.72%
Indirectly (assisters)	27	11.00%	3.70%	37.04%
Indirectly (watchers)	30	12.00%	20.00%	26.00%
Total	254	100.00%	12.99%	24.41%

Table 2
Engagement Method (shooters combined) and Self-Diagnosed Trauma (combined)

Engagement method	n	% of Self-diagnosed trauma "yes" and "maybe"
None	11	9%
Shooters (combined)	186	37%
Indirectly (assist)	27	40%
Indirectly (watch)	30	46%
Total	254	37%

In addition to the engagement method, I asked the survey participants whether they believed they had ever suffered trauma from using
an RPA to kill in combat. Understand that a self-diagnosis is not
the same as a diagnosis by a medical authority, but it does give an

indication of how these warriors feel about their experience of killing with an RPA.

As shown in Table 1, 42 percent of the overall respondents had killed directly from their platform, with another 31 percent stating that they had killed both directly and indirectly from their platform. Twenty-three percent of respondents were involved indirectly as watchers or assisters, and 4 percent of the survey respondents had no experience engaging a target either directly or indirectly. Of the 243 respondents who had killed with an RPA, ninety-five (37 percent of total respondents regardless of role) answered either "yes" or "maybe" to experiencing trauma.

What my research revealed should come as no surprise regarding the crew members who were part of a shooter platform. Of the 107 shooter respondents surveyed, fifteen of them (14 percent) answered "yes" to believing they suffered from trauma, and another thirty (28 percent) answered "maybe." Of the seventy-nine respondents who killed both directly and indirectly from their platform, ten survey respondents (12 percent) stated they believed they had suffered trauma, with another fourteen (17 percent) stating "maybe."

As shown in Table 2, combining the "directly" and "both" categories (which only occur from a shooter platform) with the "yes" and "maybe" responses, we find that sixty-nine respondents (37 percent of the total population of those in the shooter category) thought they suffered some trauma. Not to be confused with the total number of overall respondents self-diagnosing trauma, which was also 37 percent; this was just a statistical coincidence. However, of that total population, those sixty-nine shooters comprised 72 percent of all respondents who said they suffered some trauma. This same percentage corresponds to 39 percent of all shooters in the survey and 29 percent of all survey respondents. That the majority of people who stated they suffered trauma were shooters shouldn't come as a surprise. That more than one-third of the population of shooters surveyed felt they had experienced some level of trauma as a result of killing with an RPA is, however, very concerning.

Taking a look at those in the assister category, only one (roughly 4 percent) responded with "yes" to self-diagnosed trauma, and ten (37 percent) self-diagnosed as "maybe." For me this was not a surprise; you would expect lower percentages in the assister category than in the shooter category as the level of participation and ownership in the engagement process is far lower than in the shooter category. The assister marks the target, but someone else destroys it. Assisters play an important part in the overall engagement process but do not "own" the missile, bombs, or artillery used to strike a target. Nor do they have the authority to authorize a strike independently in most cases. Assisters might reasonably perceive the responsibility for the kill as being distributed among those they are working for and working with.

How "watchers" self-diagnosed for trauma was surprising and perhaps a bit counterintuitive. As mentioned previously, watchers range from anyone on an unarmed RPA crew who doesn't laze for a target to a person coordinating with the crew, such as an intelligence analyst uninvolved in flying the RPA or guiding ordnance onto the target. My survey revealed 20 percent of watchers self-diagnosed as having suffered some trauma and another 26 percent stated "maybe." This is a higher percentage than among the assisters who actually participated in the strike on the periphery and, shockingly, higher than those who actually directly struck the target themselves. This drove me to ask why.

Screen Trauma: Real to the Mind

Perhaps the reason watchers responded the way they did is found in what is called "screen trauma" or "mediated trauma." The fifth edition of the *Diagnostic and Statistical Manual of Mental Disorders* (DSM-5) is the bible used by mental health professionals in the U.S. to classify and diagnose mental disorders. According to the DSM-5, the initial criterion for a post-traumatic stress disorder (PTSD) diagnosis is a stressor, one of four exposure events:

1. Direct exposure
2. Witnessing the trauma

3. Learning that a relative or close friend was exposed to trauma
4. Indirect exposure to aversive details of the trauma, usually in the course of professional duties.

The fourth criterion of indirect exposure, according to the DSM-5, does not apply to exposure through electronic media, television, movies, or photographs, *unless* this exposure is work-related. In other words, witnessing something traumatic through a screen (or mediated trauma) is considered an exposure event if it is work-related.

This is a significant change from the DSM-5's predecessor, the DSM-IV-TR, which did not include indirect exposure to an event as a criterion for PTSD. Also, according to an American Psychiatric Association paper titled "Highlights to Changes from DSM-IV-TR to DSM-5," "the DSM-IV Criterion A2 regarding the subjective reaction to the traumatic event (e.g., 'the person's response involved intense fear, helplessness, or horror') has been eliminated." This means that being fearful for your life is no longer a necessary condition to meet the criterion of PTSD. This is important for diagnosing RPA crews who witness traumatic incidents from such a remote distance that they would have no reason to fear for their lives.

Amit Pinchevski, associate professor at the Hebrew University of Jerusalem, discussed this relationship between trauma and RPA crews in *Transmitted Wounds: Media and the Mediation of Trauma*:

The recent reference to media in DSM-V can be seen as a belated response to events that redefined the scope of trauma over the last two decades. . . .

Remote-controlled warfare gives rise to a new constellation of psychology and technology, one that fuses extreme visibility with extreme distance. . . .

Drone operator's plight captures most distinctively the stakes involved in trauma through media. However controversial, their claim to PTSD fits squarely with recent DSM-V criteria, being a traumatic media exposure suffered in the line of duty.

Providing additional evidence for mediated trauma, in 2018 a documentary film titled *The Cleaners* was released by the German film directors Moritz Riesewieck and Hans Block. The cleaners, in this situation, are thousands of Filipino workers who have been outsourced by major social media companies such as Facebook, Google, and Twitter to moderate the content posted by users of these platforms. Each cleaner looks at twenty-five thousand online photos daily to determine what stays and what goes. About every eight seconds, a cleaner views a questionable photo or post and must make a decision either to delete or ignore it. If a cleaner makes a mistake, it not only can negatively impact their job, it renders online users susceptible to the content. The implications of the cleaner's role are as valuable as they are vast, not only for the companies employing them but the countless users trafficking these public spaces on the internet.

As one cleaner states in the documentary, "I've seen hundreds of beheadings. Sometimes they're lucky that it's a just a very sharp blade that's being used to them." Cleaners see the worst of humanity and the attempt to post it online. They have witnessed online live suicides, videos of child molestation, and violent scenes of combat from around the world. This exposure to a deluge of violence, murder, sexual assault, pornography, and suicides, as you can imagine, has had a profoundly negative impact on the cleaners' lives. As if the content were not enough, there's a quantitative component to the level of trauma they witness; cleaners have a quota of twenty-five thousand objectionable photos to review daily, and if they don't make their quota, that can negatively impact their job, adding to the insidiousness of the trauma.

Due to the secretive nature of the work, there is no acknowledgment of the service the cleaners perform, no connection to the major social media companies they work for, and, most important, no mental health professionals standing by to help them deal with what they see and experience. As a result of viewing this content on a routine basis, many of the cleaners suffer from trauma. Some avoid public places because of the terror attacks they witness. Some avoid entering into intimate relationships due to the sexually abusive videos they view. Some of them, the film tells us, even die by suicide as a result of their job.

Oddly enough, the vast majority of the cleaners in the film also spoke of the importance of their job and the valuable contribution they were making in keeping the internet free of this filth. Although the cleaners provide an extreme example, it isn't hard to fathom how someone can suffer trauma through a mediated experience of events taking place half-way around the world.

The Burden of Assisting in the Decision to Kill: Intelligence Analysts

The level of responsibility and involvement that shooters and assisters have in the strike process should be obvious. What may not be obvious is the role that watchers play, particularly intelligence analysts. Although intelligence analysts aren't flying the plane or manipulating the camera, they are intimately involved in the mission, especially the decision-making process to strike a target. Prior to striking any target, the target has to meet distinct criteria known as positive identification (PID). To positively identify a target, it must be determined that it is a legitimate military target under the rules of engagement (e.g., an enemy fighter with a weapon), that it should be struck at that time to gain a military advantage, and that striking the target can be done while mitigating or eliminating collateral damage. Intelligence analysts play an important role in establishing all three of these criteria. Since a crew must have PID of the target prior to a strike, it's a guarantee that a team of intelligence analysts is involved in every strike. Analysts definitely have a certain degree of ownership of the mission and the outcome. Some may view the RPA crews as proxies acting on behalf of their decisions.

In addition to the PID process, intelligence analysts become the main focus after a strike to conduct the post-strike assessment, referred to in the military as the battle damage assessment (BDA) or bomb-hit assessment (BHA). This involves a detailed analysis of the target after it is destroyed and can, more often than not, involve some gruesome details. Several intelligence analysts I interviewed said that they were just numb from all the killing they had observed. In fact, 67 percent of the intelligence analysts surveyed said they had witnessed the killing of fifty or more enemy personnel in their career, far more than the average of any other position involved in the employment of RPAs. Given the number of respondents in my sample, this is staggering, especially considering the averages for pilots and sensor operators. For watchers, just as we saw with cleaners, the cumulative effect of observing so many incidents may take a toll on the observer over time.

The Element of Surprise: The Enemy Never Sees It Coming

There is another dynamic at play when we kill with RPAs—the element of surprise. Perhaps the most direct comparison to how we kill with an RPA is that of a sniper: lying in wait, watching the victim, prepared for the exact moment when the time is right to squeeze the trigger, without the sniper's presence ever being known.

On Killing made the point that, historically speaking, the majority of killing in battle occurs during a rout: when the enemy has turned tail and is chased down and killed from behind by a pursuing force. From a tactical vantage point, this is when it becomes easiest to kill, when the scales are tipped in favor of one combatant over the other, but this is also the psychological tipping point for the resistance to kill. When the enemy turns its back and no longer presents a physical face, psychologically it is easier to not see that enemy as another soldier, as another person, but simply as a target to be killed.

Throughout the history of warfare, humans have worked every advantage possible, manipulating both the means and the environment, to prevent a fair fight, making it more difficult to square up against your enemy and kill them face-to-face. You could argue that killing the enemy without your presence ever being known has always been the ultimate goal of warfare, especially if it can be done in a manner that prevents the attacker from bearing any risk in the situation. This is the nature of killing with an RPA and at the moment of truth, it may make it easier to kill one's enemy. In *On Killing* the aggressive predisposition of a killer was discussed and how much easier it was to kill when their presence remained undetected:

> Erich Hartmann, the World War II German ace—unquestionably the greatest fighter pilot of all time, with 351 confirmed victories—claimed that 80 percent of his victims never knew he was in the same sky with them. This claim, if accurate, provides a remarkable insight into the nature of such a killer. Like the kills of most successful snipers and fighter pilots, the vast majority of the killing done by these men were what some would call simple ambushes and back shootings. No provocation, anger, or emotion empowered *those* killings.

Every kill with an RPA is like an ambush. The enemy rarely sees it coming. From this perspective, it could be argued that the degree of trauma for the RPA crew would be somewhat reduced. It is generally a quick death, with no fear or terror on the part of the target.

The Myth: It's Easier to Kill with an RPA

To gain an understanding of how those killing remotely felt about the ease or difficulty with which they killed, I asked the survey respondents if they thought it would be easier to kill remotely than in person. Most of them had no frame of reference for killing at a distance less than extreme, but still nearly half of them said yes, that they thought it was easier to kill remotely. Those who had the context of killing in another situation, such as a manned aircraft or being physically present on the ground in the combat zone, gave the most introspective and interesting responses:

- "Different outcomes, mentally. What's easier to do remotely is harder to mentally justify. Physical combat is justified by safety of person." —USAF, Female Senior Airman, MQ-9 Sensor Operator

- "I have done both, former Apache pilot, no different."
  —USAF, Captain, MQ-9 Reaper Pilot
- "It depends on the individual. For myself I have been responsible for deaths during ground combat, and UAS kills as well. Neither bothered me much." —U.S. Army, Sergeant, RQ-7 pilot
- "There is no difference in being in an F/A-18 or other tactical jet at thirty thousand feet employing on individuals who cannot shoot back, and that of an RPA pilot doing the same from CONUS." —USMC, Colonel

Interestingly, the majority of those who have killed from both a manned aircraft and an unmanned aircraft stated that they experienced no difference. Perhaps this reveals more about the standoff nature of killing with a modern-day manned aircraft than it does about the remote nature of killing with an RPA. Regardless, it speaks to the high professionalism among our RPA warriors.

Equally interesting is the response by my study respondents who stated they killed both directly and indirectly from their own platform; this subgroup revealed during interviews that they experienced no difference in their sense of responsibility and accountability whether or not they shot from their RPA. My data analysis provided no empirical support for the engagement method as a factor that was significantly related to a negative response, suggesting other factors at work to account for more than a third of my sample self-diagnosing trauma.

The greater the quantity of violent incidents and deaths my respondents were exposed to (even indirectly mediated through a screen) equated to a higher likelihood their own mental health was negatively affected. How responsible and accountable an individual on an RPA crew felt about the outcome of a strike carried more weight than whether they struck a target directly from the RPA they were flying or whether they pulled the trigger themselves. We will explore the entire spectrum of responses to killing with an RPA in the next chapter.

Chapter Two

How Do We Respond to Killing Remotely?

The first time you blow someone away is not an insignificant event. That said, there are some assholes in the world that just need to be shot. There are hunters and there are victims. By your discipline, cunning, obedience, and alertness, you will decide if you are a hunter or a victim. It's really a hell of a lot of fun. You're gonna have a blast out here!

—General James Mattis, USMC

Is it counterintuitive to begin a discussion on how RPA crews respond to killing with a quote from a legendary Marine general and former secretary of defense that describes killing as fun? What I have discussed up to this point may have led you to believe that all RPA warriors are damaged and that their job is nothing but carnage, violence, killing, suffering, trauma, stress, fatigue, and burnout. Those descriptions, after all, dominate headlines and form the narrative within society of what we think about RPA personnel. "Drone pilot went to work, killed a terrorist, had no issues, and went home with a sense of pride and mission accomplishment" would not make for a media story that anyone would read more than once. But the reality is that RPA crews' responses to killing are as complex as any other warrior's response to killing, it just occurs in and as a result of an environment, method, and manner that we haven't observed until the last two decades. Emotional responses could be simultaneously positive and negative initially and just as equally reappraised as either positive or negative after some time has elapsed.

A person's response to killing is as unique as the situation and individual involved. There isn't a standard response. But if you talk to enough people, you start to discern some commonality among the responses. Before I discuss my research, let's look at what others have determined concerning RPA personnel's responses to killing.

An Evolution of Understanding: Research on Remote Killing

Given that RPAs have existed for the last two decades, most people have likely formed an opinion on how an RPA crew responds to killing. Many of these opinions have been expressed in journalism and presented as fact, with nothing but speculation or anecdotal information to support them. The first common narrative is that a person would be completely disengaged from the action and killing would be as emotionless as pushing a button in a video game. As this assertion goes, there should be no negative reaction to killing since there is no danger associated with the act, the killer is physically removed from the situation, and the action is mediated through a screen. According to the article "Drone-Visuality: The Psychology of Killing":

> The prevailing literature argues that the drone camera alleviates any psychological resistance to killing in war, by spatially and morally distancing the pilot from their killing and by rendering killing-by-drone a networked phenomenon.

Another prime example is a *Foreign Policy* article titled "Can Drone Operators Get PTSD?" This well-written article by a Marine infantry officer is based on the premise that there is only one way to suffer—the method we have been familiar with thus far in warfare—and extending the diagnosis of PTSD to RPA warriors somehow diminishes the diagnosis of other warriors:

> With respect to drone operators being diagnosed with PTSD, I have several concerns. First and foremost is that the PTSD diagnosis originally grew out of the experiences of veterans, rape survivors and burn victims and has always centered around a pretty basic existential question: *Did you think you were going to die?* If yes, then it's possible that you are suffering from post-traumatic stress. By extending the diagnosis to technicians watching video monitors in air-conditioned trailers outside of Las Vegas, I fear we may be diluting what has up to this point been a powerful diagnostic concept which has reduced an untold amount of human suffering.

For the record, the "fear of death" requirement for a PTSD diagnosis was dropped in the DSM-5, as previously discussed, but that brings us to the second common narrative. It occupies the other end of the spectrum and portrays all RPA crews as damaged and suffering from PTSD as a result of some sort of moral injury for killing mercilessly without honor and against one's personal values.

Fortunately, our understanding of the human response to killing remotely has evolved over time as the library of research has grown. U.S. Air Force psychologist Dr. Wayne Chappelle has been at the forefront of research that has formed the basis of our factual and empirical understanding thus far of how RPA personnel respond to killing. It's therefore fitting that we review what Chappelle and his associates have determined through nearly a decade of research and observation.

The first time PTSD was mentioned in connection with RPAs was in a 2012 study by Dr. Chappelle and other specialists titled "Prevalence of High Emotional Distress and Symptoms of Post-Traumatic Stress Disorder in U.S. Air Force Active Duty Remotely Piloted Aircraft Operators." At the time of the study, most of the information concerning RPAs and the human response to remote killing was along the two narratives described above. Here's what the study concluded:

> Participants included 670 USAF RPA Predator/Reaper operators and 751 noncombatant airmen . . . The most commonly cited stressors among RPA operators included long hours, shift work, deployed in-garrison status, ergonomic design of the ground control station, and sustaining vigilance to large amounts of real-time visual and auditory data. Combat-related stressors were not rated as top sources of stress. Rates of clinical distress and PTSD were higher among RPA operators (20% and 5%, respectively) in comparison to non-RPA airmen (11% and 2%, respectively).

While the 2012 study exposed RPA stressors, it wasn't until Chappelle's 2014 study titled "An Analysis of Post-Traumatic Stress Symptoms in United States Air Force Drone Operators" that we first got a glimpse of PTSD's existence in the RPA community. Dr. Chappelle produced the first scientific study that indicated all was not well.

> Given there is a paucity of empirical research assessing drone operators, the purpose of this study was to assess for the prevalence of PTSD symptoms among this cohort. Of the 1084 United States Air Force (USAF) drone operators that participated, a total of 4.3% endorsed a pattern of symptoms of moderate to extreme level of severity meeting criteria outlined in the Diagnostic and Statistical Manual of Mental Disorders–4th edition. The incidence of PTSD among USAF drone operators in this study was lower than rates of PTSD (10–18%) among military personnel returning from deployment but higher than incidence rates (less than 1%) of USAF drone operators reported in electronic medical records.

Chappelle and his colleagues published two additional studies in 2014, "Symptoms of Psychological Distress and Post-Traumatic Stress Disorder in United States Air Force 'Drone' Operators" in July and "Assessment of Occupational Burnout in United States Air Force Predator/Reaper

'Drone' Operators" in August. Both were results of a web-based survey of 1,084 RPA aircrew. The July publication revealed:

> ...the most problematic self-reported stressors are operational: low manning, extra duties/administrative tasks, rotating shift work, and long hours. The results also reveal 10.72% of operators self-reported experiencing high levels of distress and 1.57% reported high levels of PTSD symptomology. The results are lower than findings from the 2010 survey and from soldiers returning from Iraq and Afghanistan.

The August article looked at those surveyed through the lens of burnout and determined that RPA crews experienced high levels of exhaustion (20 percent) and high levels of cynicism (11 percent), while only 3 percent reported low levels of professional efficacy.

In 2015, Chappelle researched psychological distress and PTSD among Air Force intelligence personnel typically associated with supporting RPA operations at Distributed Common Ground Systems (DCGS). The studies yielded results similar to those from previous studies of RPA aircrew: 14.3 percent self-reported high levels of psychological distress while 4.22 percent met "provisional diagnostic criteria for PTSD."

In 2017, in addition to occupational stressors and burnout, Chappelle added predictors of suicide ideation to the study and determined that suicide ideation was self-reported by 5–6 percent of the 4,340 remote warriors surveyed (27.8 percent of which were RPA aircrew, 60.3 percent intelligence personnel, and 11.9 percent cyber warriors).

In the 2018 study "Emotional Reactions to Killing in Remotely Piloted Aircraft Crewmembers During and Following Weapon Strikes," Chappelle looked at factors that might cause negative emotional responses to killing remotely:

> The purpose of this study is to gather both quantitative and qualitative data on the emotional reactions of remote warriors and examine potential occupational (e.g., number of years as an RPA crewmember, prior military experience, prior combat deployments, and total number

of weapon-strike missions), demographic (i.e., age, marital status, gender, and dependents living at home), and mission-specific (i.e., target familiarity, mission outcome, and high-definition vs. standard-definition video feed) correlates of negative reactions.

The 2018 study determined that witnessing civilian casualties put RPA crews at elevated risk for negative reactions to killing.

In a February 2019 study, "Combat and Operational Risk Factors for Post-Traumatic Stress Disorder Symptom Criteria Among United States Air Force Remotely Piloted Aircraft 'Drone' Warfighters," Chappelle conducted another online survey of 715 RPA warriors with firsthand exposure to killing remotely. He found that 6.15 percent met the symptom criteria for PTSD, of which those in the age ranges thirty-one to thirty-five and thirty-six to forty, and those who worked fifty-one hours or more per week demonstrated greater possibility of having PTSD symptoms. "The results of this study suggest that specific types of exposure and participation in missions with specific outcomes, albeit via electronic, remote means, are associated with an increased risk for meeting PTSD symptom criteria."

This is just the beginning of the research needed to arrive at conclusive findings. But what we can glean from this research is that the vast majority of RPA crew do not get PTSD, and the major cause of their stress can be factors other than killing, such as sleep problems, fatigue, and burnout.

A Spectrum of Responses to Killing Remotely

The reality is that after nearly two decades, the human response to killing remotely is still relatively uncharted territory. Responses to killing remotely are as complex and varied as other warriors' responses to killing. My research was both quantitative and qualitative and consisted of surveying 254 RPA and intelligence personnel and conducting more than fifty in-person interviews. The chart on the next page depicts responses to my request that participants describe any and all psychological responses to killing remotely.

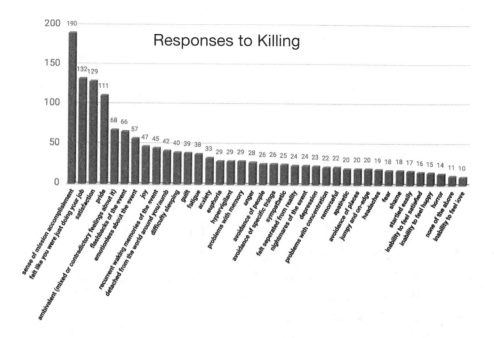

Since I was working with a community of professional warriors, I was not surprised to find that positive feelings such as a sense of mission accomplishment, doing one's job, satisfaction, pride, and joy topped the chart. Neutral feelings such as ambivalence or being emotionless about the event were the next dominant group. The remainder of the chart focused on the most common negative responses to killing. Nearly 26 percent of those surveyed stated they experienced flashbacks of the event, 17 percent stated they had recurrent waking memories of the event, 16 percent stated they felt detached or numb, and 15 percent admitted to having difficulty sleeping. How long these negative responses persist determines whether these initial negative responses result in lasting trauma.

The First Kill: Usually the Most Difficult

As one might expect, the first time a human kills another human tends to be a highly difficult experience and nearly always elicits either a positive or negative psychological response. Subsequent kills are often reported to occur with less impact on the killer. Air Force Colonel Joseph Campo studied this extensively by interviewing 110 RPA aircrew and published the results in an article titled "Distance in War: The Experience of MQ-1 and MQ-9 Aircrew." Campo concluded that:

MQ-1/9 aircrew displayed relatively high rates of emotional response to their first kill, with nearly three-quarters of interview subjects reporting a first-strike emotional response. Demographic variances, including prior-manned-aircraft experience or prior-combat deployments, failed to demonstrate any statistically significant differences in emotional response rates to killing via MQ-1/9. Stated another way, the emotional responses to killing via RPA did not vary based on whether the aircrew previously flew an F-16, deployed as a security-forces airman, or has never deployed or flown a manned aircraft.

The most common response was a positive emotion following mission success or supporting friendly ground forces juxtaposed with negative emotions for the taking of human life. Inter-strike, aircrew emotions were highly dependent on the details of a particular mission, with specific focus on the safety and success of friendly ground forces and the avoidance of civilian casualties and collateral damage. The status, and specifically safety, of friendly ground forces was found to result in the highest rates of emotional response across the entire study. . . . Combining the response rates across the emotional, social, and cognitive domains resulted in a first-kill psychological response rate of 94% for all aircrew.

My interviews revealed the same conclusions as Campo's; the first kill is an emotional mess. I interviewed one Reaper pilot just three weeks after his first shot. The details and his responses to killing were still freshly etched in his memory and illustrative of common first-shot experiences of RPA personnel:

Prior to my first shot, I was curious about the experience and how I would react. Throughout the experience I was consciously trying to remind myself what I was doing despite how stressed out I was at the moment. I had been flying for six months so far and hadn't shot yet. I wanted to remember how it felt the first time.

Then one day it happened, and it happened so fast I could barely process it. To summarize the event, there were some ISIS fighters trying to cross a river in a boat to reach a town. The concern from the JTAC was that we would lose them if they got across the river and disappeared into the city. So we had to strike and strike fast. I remember spinning up for the strike being very nervous. In my mind, I knew what I was supposed to do, but it was ten times harder when my heart was beating 190 beats per minute.

Prior to every shot, we go through a missile briefing to get it ready to go. It's a pretty lengthy brief, and I remember not being able to get through it. I kept getting stopped up on the first step. I think my body was in a fight-or-flight mode. I had executed the brief numerous times in training without difficulty, but I think my body knew this time it was real. To calm me down, a more experienced pilot came into the box to help me get through the brief. I was sweating profusely, my hands were shaking, and my butt was shaking in the seat. I remember the other pilot asking, "Are you ready?" I said "yes," not really knowing what else to say. The JTAC, in this burly man-voice, said, "TOT [time on target] immediate!" The JTAC said it so casually, it made me think he had done this several times and here it was my first time. I just kept thinking to myself, don't mess it up.

We pushed the aircraft in, got a cleared hot from the JTAC, and rifled the missile. Simultaneously, the easiest and hardest thing in the world to do at that moment was to push those two little buttons to launch a missile. I saw a flash as the missile left the rail and before I knew it, I saw an explosion over the boat. Then we switched the camera from infrared to electro-optical [daytime color camera] and at that moment I realized what I had done. The boat was on fire, a few bodies were floating in the water, and the water was turning red.

I don't think my mind knew how to deal with it since the experience was so intense and something I had never felt before. It was like combining feelings of fear, a tingly kind of feeling like having a crush on somebody, and a feeling of guilt. It felt like hitting an emotional funny bone. I remember trying to hold on to that moment to try to process it. But it escaped me. I could see how somebody would throw up after that experience. It was an alarming moment laced with complex unfamiliar emotions clashing together simultaneously. At the core of the emotions was this feeling that I had done wrong to those people, like when you accidentally hurt someone playing sports and you feel like an asshole for it. I felt like an asshole.

In the next thirty minutes, I went on to shoot a building with IEDs in it, shoot another enemy fighter, and then I buddy lazed for three other shots. The whole duration of it all was about forty minutes, but it felt like it only took ten minutes. After the strike, when I wrote my after-action report, I could hardly remember any details. I had to ask the more experienced pilot to help me recall the events.

Physiological Responses: What the Mind Perceives the Body Believes

As further evidence of an RPA crew's connectedness to the mission, when killing remotely they experience a wide spectrum of physical responses as well. This may seem counterintuitive since they are killing from such great distance, but the sensations are remarkably similar to those felt by warriors who square off face-to-face with the enemy. What this suggests is that it isn't fear that elicits these responses, but rather the backlash from the mind's anticipation of overcoming the resistance to killing.

Nearly every strike begins with the crew experiencing an adrenaline spike and an elevated heart rate almost immediately after being told to "stand by to receive a nine-line." The effects of that adrenaline rush and elevated heart rate start to impact performance right away. Something as simple as writing the information needed for the strike received from a JTAC over the radio can become difficult during this period. For those who have mastered their body's responses to killing with RPA, this may be the only physiological response they still experience. Interviews revealed that this response was the most common among RPA warriors, even for individuals who had conducted more than fifty strikes.

As discussed in *On Combat*, once the heart rate gets up around 170 beats per minute the brain is in fight-or-flight mode and fine motor skills become diminished. This can be very problematic at the moment of truth, since fine motor skills are required to launch a missile and to keep the laser on the target for thirty seconds. Fortunately, very few RPA crew members revealed this as a problem and hardly anyone spoke of having issues with fine motor skills after their first strike. This is indicative of the muscle memory formed through extensive training and an aviator's ability to compartmentalize distractions during a strike. In other words, the same initial response, tempered by training, that all warriors experience.

How the crew responds beyond the initial adrenaline rush really differs by individual. What is interesting to note is that the wide range of physiological responses is comparable to the range of responses experienced by people involved in face-to-face lethal force encounters. I heard reports of sensor operators getting tunnel vision, their sight limited to the target on the screen appearing in laser-like focus, while the edges of the screen grew blurry.

Auditory exclusion was revealed in several interviews, predominantly from sensor operators who stated they were so focused on the target that they could no longer hear the pilot counting down as the missile got closer to impact. Temporal distortion was also frequently mentioned. The time it takes for the missile to reach its target after launch seems like an eternity for most, whereas an extremely busy period of a mission may appear to take only a few minutes, when in actuality it took four times that. Some reported cold hands, uncontrollably shaky hands, shaky legs, and vomiting after a strike.

Memory is also affected in a manner comparable to the experience of those who have engaged in deadly force, face-to-face encounters. And consistent with the survivors of these face-to-face encounters, many RPA warriors spoke of remembering very vivid details of those they killed, while others spoke of being able to remember hardly any details at all.

One Predator pilot named David spoke of a memory gap in which skills he had honed over his career were suddenly lost.

During one particular stint, I was given nine nine-lines [a request for a strike] in one week. It was so strange and bizarre that crew members assigned to my missions were spooked to fly with me.

Temporal distortion was commonplace. Night sweats, dread of work, and depression were the norm. Our families knew something was up. It also took a toll on them, too. They didn't have to be told when a missile was employed. A prolonged shift after a strike made it obvious. If the events of a lethal day weren't hard enough, the continual analysis and review we made of the strike to improve our skills and techniques for the next strike drove events deep into our memory.

There are certain things I encountered that to this very day remain some of the most vivid things that have ever happened to me. They remain crystal clear in my mind. One particular event stands out: a situation I observed that ended with civilian casualties, as a result of an Apache helicopter attack of a target. After witnessing this and taking a short break to gain my composure, I got back in the seat to continue on with a mission, but I couldn't remember how to move the aircraft. Whatever mental protection my mind had in wiring kicked in. My particular coping mechanism is that I forget things if I am under an extreme amount of stress. In that situation, I had completely forgotten how to fly.

I believe there is a very real tie to what we see with our eyes and our body's response to it. After every strike, my body would completely sweat while sleeping that evening. I had never experienced that before or after. It was like my body was trying to respond to something that it didn't know how to deal with. Those experiences are now part of me, part of who I am, and they shape you.

It is hard to overcome your natural tendency to not want to kill another human. It must be painful to kill, it must be injurious, it must always be difficult. It should never be easy to put someone in the line of fire or we lose our humanity. In many cases, your back is against the wall. You must pull the trigger. In those times when it's self-defense of troops on the ground it is much easier. Those long loiter times doing a POL gives you a bizarre sense. When I returned to Afghanistan as a manned pilot, I flew over a lot of the places that I had flown over as an RPA pilot and it was a surreal experience. I felt as though I had been there. I was reliving the hours on end circling overhead with a camera. It was almost a feeling of déjà vu.

The true impact of killing on this generation of operators will not be realized for some time. But it will come out. These men and women are patriots and did what their nation asked them to do. But they are misunderstood and often those who have snapped try to sensationalize how illegal things were. Nothing about it was illegal, and perhaps that was the hardest thing of all to understand and reconcile.

Misunderstood patriots, indeed. Next, we'll look at the demographics of who is affected by these experiences the most.

Chapter Three

Who Is Most Affected by Killing Remotely?: The Demographics

> When plans are being made for operations in these environments, it is usual to consider only two possibilities: either placing a machine in the environment or placing a protected man there. A third possibility, however, would in many cases give more satisfactory results than either of the others. This possibility employs a vehicle operating in the hostile environment under remote control by a man in a safe environment.
>
> —Engineer John W. Clark, "Remote Control in Hostile Environments," 1964

Now that we have an understanding of how RPA crew members respond to killing, the next logical questions are: Who is most affected and why? While my research was aimed at finding answers to these questions, a plethora of variables—notably, the diversity of RPAs employed across different branches of the U.S. military—made trend identification and isolation of causal factors challenging and even conflicting at times.

Underlying my research methodology were two essential criteria for evaluation: those who had used an RPA to kill and how they responded to killing. From that common ground I endeavored to spot trends and correlate the demographics of individuals who had the worst psychological response to killing remotely. The interviews I conducted with RPA personnel added depth to the surveys, reinforced my findings, and in some instances revealed other trends not apparent from quantitative analysis.

Finally, my undertaking was the first comprehensive data collection in the RPA community at large. Dr. Chappelle's work described in the previous chapter is an informative starting position, particularly when it comes to how Air Force RPA crews respond to killing. However, his research is limited in scope to only the U.S. Air Force. It does not take into consideration the U.S. Army, U.S. Marine Corps, the intelligence community outside of the Air Force, and the Royal Air Force RPA personnel who have employed RPAs with lethal effects for quite some time. I wanted to know if the conclusions determined by Dr. Chappelle were universally applicable across the entire RPA community, or if there were differences. And if so, why? To my knowledge, no one has studied the RPA community in its entirety across the services and across all RPA employed. This effort brings to light trends related to killing with RPA that up to this point may have flown beneath the radar.

Not Your Average Joe: An Extreme Profile of a Remote Killer

Let's be honest, statistics are not very exciting to most of us. Fortunately, we are less concerned about the statistics and more concerned about what they reveal. To represent the results of my research, I'll personify the most negative trends across each demographic through a profile of a fictional person named Joe. Joe represents the demographics that have the most psychologically challenging experience of killing with RPA across all the variables. Joe is not the average, or even a good, representation of how the overall community responds to killing. Instead, he represents an extreme minority. The average RPA warrior, just like the average traditional warrior, doesn't experience the same psychological challenges to killing as Joe.

To understand which populations are most affected, I focused on the demographics of age, sex, education, service, crew position, aircraft flown, number of hours flown, method of engagement (direct or indirect), number of engagements, and the estimated number of enemy combatants killed with RPA in combination with qualitative factors derived from survey questions. These qualitative factors include whether the participants thought their training mentally prepared them for killing, whether they had a post-killing discussion with a mental health professional, their job satisfaction, events that occurred during missions, and varying amalgamations of the aforementioned questions. Each demographic and qualitative question was assessed against a series of questions: self-reported PTSD, prevalence of at least one negative response (a PTSD symptom from the DSM-5), prevalence of multiple negative responses to killing, and rationale for self-reporting PTSD. Joe represents the population of each demographic with the highest rate of identified challenges to killing remotely. Let's meet Joe.

Male in His Early Thirties

Joe is a male between thirty-one and thirty-five years old. Although there was no significant difference between Joe's responses and those of the average female respondent, as a male Joe is representative of 95 percent of those surveyed, and the male-dominated RPA community. Joe's age is congruent with Dr. Chappelle's research as well. In the early thirties age group, an officer or enlisted service member is at the peak of his tactical experience. He has served long enough to master the tactical and technical aspects of the trade, but not so long that he has moved into a job where he is tied to a desk more often than flying. Traditionally this is also the age group in which aircrew become instructors and teach others the tactical nature of fighting, causing them to spend more time in the cockpit. As a result, this age group is very experienced, leaned on heavily by their leadership for the execution of flight operations, and has more opportunities to be exposed to events during a mission that may negatively impact crew members.

Armed

Joe serves in either the Army or the Air Force. This is an interesting observation since the two services differ significantly in RPA employment method, crew composition, training, and culture. Air Force RPA crews employ from the U.S. whereas Army crews forward deploy. The Army RPA crew consists of enlisted soldiers while the Air Force crew is comprised of an officer and an enlisted airman. Training and cultural differences were previously discussed in Section I, Chapter 4. The one constant between the two services is that they both employ an armed RPA such as the MQ-9 Reaper (U.S. Air Force) or the MQ-1C Gray Eagle (U.S. Army). This makes a significant difference. In fact, Joe is twice as likely to experience a negative psychological response to killing than those who don't employ ordnance directly from the RPAs they fly. So Joe is a crew member of an armed RPA.

Married

Joe is married and has children. Interviews determined that those married with children were more prone to recognize the humanity of their targets, which led to a more negative response to killing. When Joe has a bad day at work, he most likely chooses not to share any information with his spouse when he gets home. He doesn't want to burden his spouse with gory details of his work, and he is uncertain about the information he can share due to the confidential nature of his job. In contrast, those RPA personnel who disclosed they did share information with their spouse about their work had a better support structure and were less likely to self-report feelings of PTSD.

Exhausted

Joe works more than fifty hours a week and feels exhausted most of the time. Although this isn't a cause of negative responses or PTSD, we know that exhaustion and cumulative fatigue can chip away at a person's mental armor and make him more susceptible to registering a negative response to a traumatic event. Feelings of exhaustion may manifest in being short-tempered and impatient with family or friends and mildly disgruntled with the job. Working more than fifty hours a week also socially isolates Joe since he is too tired for social engagement with peers in his time off. Social interaction with peers, as we know, is important for developing a support structure and relieving stress.

Educated

Joe has attended some college but does not have a degree. This indicates that Joe is most likely an enlisted service member since officers traditionally have a bachelor's degree prior to entering the service. Level of education, however, was not a significant indicator of a person's responses to killing remotely, so Joe could just as easily be an officer with a master's degree, which was a close second education demographic for those who experienced negative responses to killing.

Joe is most likely a sensor operator. Surprisingly, the crew position was also not a consequential discriminator. Sensor operators had the most negative responses, but pilots and intelligence analysts were not far behind. I attribute the insignificant differences in crew position responses to the distribution of responsibility among crew members; more on that in the chapter on group absolution.

Experienced

While crew position and education aren't noteworthy, what is significantly relevant is that Joe has conducted more than fifty strikes resulting in the death of fifty or more enemy combatants. The chart below depicts the distribution of enemy combatants killed by my respondents.

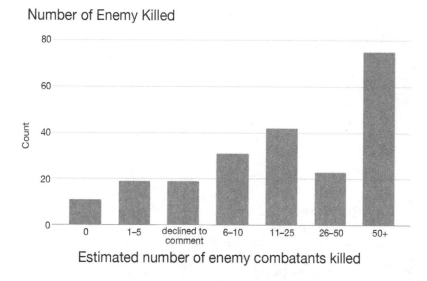

Estimated number of enemy combatants killed

Fifty-four percent of Joe's peers who had killed more than fifty people self-reported feelings of trauma, 81 percent of them had at least one negative response to killing, and 68 percent of them had multiple negative responses to killing. However, it's not just the sheer number of strikes or enemy killed that prompt negative responses or feelings of trauma. The volume of strikes participated in significantly increases the odds of a person having more exposure to the operational scenarios that actually cause psychological issues.

The majority of Joe's strikes have been directly from his own platform, but when he lazes for another aircraft to destroy a target it does not affect him any differently than if the bomb or missile came off his own aircraft.

So those who killed the most were the most likely to have difficulty. But don't forget that this is only a minority of the overall group, who did not have problems.

Mentally Unprepared

The first time Joe killed with an RPA he had a negative response to it. He did not feel as though his training mentally prepared him for the moment. This was a common response.

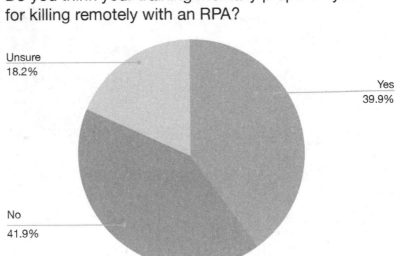

Do you think your training mentally prepared you for killing remotely with an RPA?

Unsure
18.2%

Yes
39.9%

No
41.9%

Even though help was available, Joe chose not to discuss the negative response to killing he experienced after his first strike with a psychologist or chaplain for fear that others would find out or that he might be pulled off of flight status. A removal from flight status would let down Joe's fellow warriors, who would have to work harder to cover his absence.

The U.S. Air Force and British Royal Air Force have by far the best participation in post-killing discussions with a mental health professional or chaplain with 44 percent of participants from those services stating that they had talked to someone post-strike as opposed to 5 percent for the Army and Marine Corps. This is attributed to the implementation of programs by the USAF and RAF such as Human Performance Teams (HPTs) and Trauma Risk Management (TRiM).

In the Air Force, HPTs consist of mental health professionals and chaplains embedded directly in RPA units. These professionals also have the same level of security clearance, so the operational details, which may be the part of the mission most disturbing to a crew, can be discussed. It is a fantastic program, but helpful only if used.

TRiM began in the Royal Marines and has expanded to all of the British Armed Forces as a method of identifying possible traumatic events that occur during a mission and then allocating resources, particularly peer support, to assist an individual in coping with the aftermath.

Even with programs such as these in place, it is apparent that there is still room for improvement, as depicted in the chart below. The need to improve is all the more urgent since there is good cause to believe that programs such as HPT and TRiM can have a positive effect on the well-being of RPA crews.

Following a mission where an RPA was used to kill, is it common practice in your service for the aircrew to discuss the event with a mental health professional or anyone else?

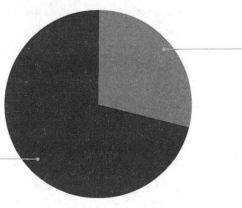

Yes
28.7%

No
71.3%

The Three Events Most Likely to Cause a Negative Response

Subsequent kills after Joe's first strike produced fewer negative psychological responses, unless he experienced one of three events during a mission: witnessing friendly forces being killed or wounded, civilian casualties occurring as a result of a strike he conducted, or killing a person after spending a long time developing a pattern of life on them.

When Joe witnessed friendly forces die during a mission and couldn't do anything to prevent it from happening, he felt like a failure. It was a feeling of letting his fellow service members down, despite never having set foot in the same country and never meeting those friendly forces. He and 24 percent of his fellow warriors in the same age group had multiple PTSD symptomatic responses following these missions, including some flashbacks of the event at random and often most inconvenient times. These flashbacks are mostly a replay of the situation over and over in his mind, trying to figure out how he could have done something different that would have led to a successful outcome. This event has also led to some difficulty sleeping, which is further exacerbated by Joe's rotating shift schedule.

If Joe knew that civilian casualties occurred as a result of a strike he conducted, he and 11 percent of his fellow warriors in the same age group had multiple PTSD symptomatic responses. This is indicative of the level of professionalism of the aircrew, caution that is taken to prevent this tragedy, and how emotionally devastating the experience can be.

When Joe killed a target after conducting a long pattern of life of the individual, he and 16 percent of his fellow warriors experienced PTSD symptomatic responses. This reflects Joe's connection to the mission and his acknowledgment of the humanity of the enemy combatants he kills. This was also the leading response to self-reporting feelings of PTSD. This topic is discussed further in the chapters on intimacy with the target and target attractiveness.

The remote nature of killing with RPA is not always cognitively easy to grasp or rationalize. In fact, Joe and 21 percent of his fellow warriors had multiple negative responses to killing remotely because they said that it seemed surreal even though they knew that it occurred. I'll discuss this more in the chapter on distance.

Joe thinks he might be suffering PTSD from one of these operational scenarios but is uncertain. In fact, 13 percent of all RPA warriors surveyed said they thought they were suffering PTSD while another 24 percent said they might be suffering PTSD but were unsure.

Do you think you have ever suffered from PTSD from using a UAS/RPA in a combat situation?

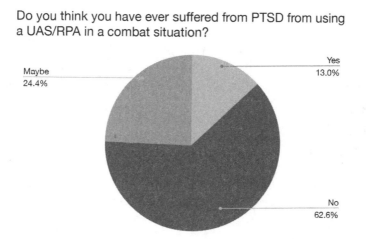

Maybe
24.4%

Yes
13.0%

No
62.6%

These self-reports are not a PTSD diagnosis, but indicative of the challenges that RPA personnel encounter. In 2019, Dr. Chapelle determined from his research that 6 percent of USAF RPA personnel demonstrated PTSD symptoms, a 2 percent increase from his previous research. Again, these numbers are very small, and the expectation created by the media is that greater numbers than this will have PTSD. Across the years and throughout the armed forces (not just among RPA crews), there are many veterans who think there is something wrong with them *because there is nothing wrong with them!*

Interestingly, that 2 percent increase occurred even after the implementation of the Human Performance Teams in Air Force RPA units. It's entirely possible that Chappelle's respondents underreported or that my respondents overreported. Regardless, both results are disturbing for the community. One variable between the two studies is that Chappelle surveyed aircrew in operational RPA squadrons, whereas my research was not limited to current active-duty members. A possible cause of the disparity in numbers is that my respondents had more time to reflect on what they had experienced after leaving the service. Another possible disparity is that the introduction of programs such as HPT and TRiM have provided the much-needed assistance to those struggling with the effects of killing remotely.

The most common operational scenario that caused RPA crew members to self-report feelings of PTSD was killing an individual after observing their humanity during a long POL mission. The second most probable scenario was witnessing friendly forces being killed or wounded and being unable to prevent it. The third most probable operational scenario was that killing remotely seemed surreal. Civilian casualties were fourth in order of responses.

Explanation of why the respondent answered yes or no to self-reported PTSD

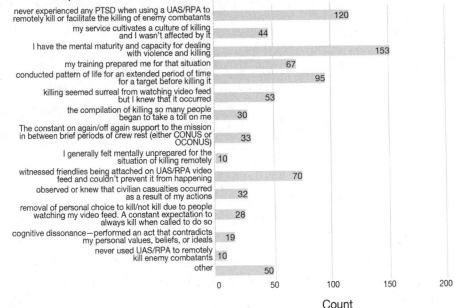

	Count
never experienced any PTSD when using a UAS/RPA to remotely kill or facilitate the killing of enemy combatants	120
my service cultivates a culture of killing and I wasn't affected by it	44
I have the mental maturity and capacity for dealing with violence and killing	153
my training prepared me for that situation	67
conducted pattern of life for an extended period of time for a target before killing it	95
killing seemed surreal from watching video feed but I knew that it occurred	53
the compilation of killing so many people began to take a toll on me	30
The constant on again/off again support to the mission in between brief periods of crew rest (either CONUS or OCONUS)	33
I generally felt mentally unprepared for the situation of killing remotely	10
witnessed friendlies being attached on UAS/RPA video feed and couldn't prevent it from happening	70
observed or knew that civilian casualties occurred as a result of my actions	32
removal of personal choice to kill/not kill due to people watching my video feed. A constant expectation to always kill when called to do so	28
cognitive dissonance—performed an act that contradicts my personal values, beliefs, or ideals	19
never used UAS/RPA to remotely kill enemy combatants	10
other	50

Satisfied with the Job

Despite all of this, Joe is very satisfied with the work he is doing, as depicted in the chart below. He thinks his job is extremely important and that he is making a difference. Job satisfaction aside, Joe does feel isolated and misunderstood by his fellow service members outside the RPA community. He also detests the portrayal of RPAs in the media; I'll discuss this in the chapter on culture.

Job Satisfaction

neither satisfied nor
9.1%

very dissatisfied
4.5%

somewhat dissatisfied
5.0%

somewhat satisfied
35.0%

very satisfied
46.4%

RPA warriors are similar to other warriors in how and why they respond to killing, which may be surprising given the remote nature of the job. Most traditional warriors do not experience any long-lasting effects of participating in what is quite possibly the most challenging thing a human can undergo—taking a life. The majority of RPA warriors respond in kind. The extreme minority who do have a resultant negative experience are older, experienced males who are exhausted, fly an armed RPA, and experience a specific traumatic operational event during a mission. In the next section I take an in-depth look at how those events shape the experience of the RPA community and contribute to the environment they operate in.

SECTION III

All Topics Considered

Chapter One

Are We at War?

Men do not fight because they have arms. They have arms because they deem it necessary to fight.

—Hans Morgenthau

And over time, as codes of law sought to control violence within groups, so did philosophers and clerics and statesmen seek to regulate the destructive power of war. The concept of a "just war" emerged, suggesting that war is justified only when certain conditions were met: if it is waged as a *last resort* and in self-defense; if the force used is proportional; and if, whenever possible, civilians are spared from violence.

—President Barack Obama in his 2009 Nobel Peace Prize
acceptance speech

For a warrior, war is what sanctions the killing of another human being from a legal, ethical, and moral standpoint. Killing outside the bounds of war is widely considered murder; therefore, the permission to kill under the banner of war is vitally important to the warrior's process of rationalizing what he is doing for his country.

Since 2001, RPAs have been employed throughout the world, in numerous countries, and during varying degrees of conflict. This has led to a popular (and incorrect) perception that RPA crews are suffering from moral injuries because they are not copresent where they fight and because they can kill with impunity in any geographical area where the United States is engaged in conflict; that they are killing without being "at war." In other words, that RPA crews are murderers. There have also been incorrect assertions that RPAs have changed the nature of war into something akin to remote control assassinations. So I ask, within the context of RPAs, "Are we at war?" and "Have RPAs changed the nature of war?" To answer these questions, we need to start with the writings of a Prussian military theorist on the definition of *war*.

War as Defined by a Dead Prussian

Carl von Clausewitz's theories on war are widely accepted as dogma among the vast majority of military scholars. According to Clausewitz in *On War*, "War is the continuation of politics by other means." He describes war as a duel between two sides and goes on to define war as "an act of force to compel our enemy to do our will." Clausewitz stated that this is the objective and nature of war and that it is enduring and does not change. What changes is the character of war; how it is executed and the weapons used, otherwise known as the "means" of war. Clausewitz recognized that it is the nature of humans in civilizations to make advancements in technology and improvements in the weapons used in war, but ultimately the duel between two sides, and the constant struggle to bend the enemy to our will, does not change. So the means may change, but the objective remains firm. It is the nature of war.

Now, while Clausewitz's definitions are commonly accepted as maxim among military scholars, they are not indisputable. Regardless of which side of the debate one falls on, it is safe to say that killing someone from seven thousand miles away at the tactical level of war is a relatively new development in warfare and we are just now starting to understand it. But I hold firmly in the Clausewitz camp that RPAs have evolved the character or means of war, but not the nature or objective of it. Interestingly, that is not the consensus among those I surveyed. Half of those 254 RPA personnel I surveyed stated they thought that RPA had actually changed the nature of war. Even more interesting were their own responses to the survey question "Do you think RPAs have changed the nature of war or simply improved the tools by which war is fought?" First some replies from those who think RPAs *have* changed the nature of war:

- Yes, violence is too easy. Politicians feel they don't have to consider any negative implications because they feel there are none since RPA crews are removed from danger.
- I believe that the use of RPAs has become the "easy" answer for policy makers. The lower risk to forces makes an easy sell to utilize the RPA versus putting actual troops in harm's way and as such, we [as a society] in America have become numb to the "drone strike" news reports to the point that most don't even register in the consciousness of the population anymore.
- Changed the nature; the barriers to kinetic operations are decreased through remote operations.
- It has changed warfare as it has removed one of the safeguards from politicians' decision-making, which is the loss of American lives to conduct an operation.
- Yes, we no longer have to risk friendly human life to engage our adversaries.
- Yes. Command of life and death are now controlled daily from a remote location. Over time, the gravity of the situation is lost and taking life becomes less meaningful.
- Both, we have become disconnected from war utilizing UAS/RPAs.
- UAS have changed the nature of warfare by emotionally

153

distancing the UAS operators from the act of killing regardless of the aftereffects of the action.

- Yes. RPA have allowed individuals to remove themselves and in turn hold themselves less accountable for the actions in which they were fulfilling their job.
- I think they absolutely changed the nature of war. Although at its core it remains a clash of human wills, and true to its ancient nature, my sense is that unmanned and autonomous technology has introduced a new and complex ethical dimension to the decision-making aspect of killing in combat.
- It's changed it entirely, but we haven't seen the capabilities implemented even close to fully yet, [such as] armed swarms.

It's obvious that those who think that RPAs have changed the nature of war don't think it is for the better and their sentiments nearly reflect those of the anti-drone crowd: RPAs make it easier for politicians to decide to go to war, RPA personnel are emotionally distant from killing, less accountable for their actions due to their physical distance, and therefore the act of killing becomes too easy. This line of thought also views RPAs as the first step in what will eventually evolve into a morally and ethically questionable autonomous killer robot.

These aren't monolithic sentiments for the RPA community. The opposite end of the spectrum, the "Clausewitzians" who think the nature of war is enduring and that RPAs simply improved the tools of warfare, is equally represented:

- Nature of war, as defined by dead Carl, has not changed—it is still an act of violence to compel an enemy. Being ten thousand miles away is no different from being ten thousand feet above except that the distance makes the operator safer.
- Improved the tools. Ground personnel will continue to be needed to meet operational and strategic objectives.
- UAS are just tools. Nothing more than that. The only difference between using them as opposed to an artillery piece is that we can actually see the impact of the round/missile.

- RPAs are evolutionary, not revolutionary. Humans have striven to increase "distance" in all forms of warfare; the RPA increases physical distance (immediate risk) while shrinking emotional distance (due to the HD/zoom levels and time spent observing patterns of life development). While this is new, it certainly hasn't changed the nature of war, which is the destruction of people and things for political gain.

- RPAs are the descendants of the sling, the archer, the mortar, and the high-altitude bomber. They are the natural evolution of war strategy, which is to gain advantage without vulnerability.

- It gives the illusion that it's not real war to those on the outside. To those flying it and its targets the results are the same as dropping a bomb from thirty thousand feet.

- Improved the tools by which war is executed; the remotely piloted aircraft carrying ordnance is little different from a "man-in-the-loop" piece of ordnance dropped from a manned aircraft.

- Not the fundamental nature of war. "Drone strike" is a misnomer, as we have not truly automatized actual strike authority, weapons release, or guidance. Realistically, a drone strike is just an air strike with some sort of remote observation or weapons delivery.

- Simply changed the tools. Having operated both manned and unmanned aircraft in the same theater with very similar mission sets, I believe it's merely a difference in tools.

- The nature of war does not change. However, the character of war has undoubtedly changed due to unmanned technology and reach back/remote operations.

To those who expressed that RPAs have simply improved a tool of warfare, their responses reflected the more positive aspects of using RPAs in warfare; they remove some people from harm's way but don't replace the need for ground troops, RPA employment is similar to the use of other weapons such as strike aircraft from a tactical and practical standpoint, and the human is still in the lethal decision-making loop (i.e., RPAs are not lethal autonomous weapons deciding to kill without a human's input and control).

So even among those who employ RPAs, it is clear that there is much discord on the impact of this weapon as it relates to the nature and character of war. Perhaps we should accept that this relatively new tool of warfare still adheres to the nature of war and just changes the character of war as did every new evolutionary weapon that preceded it as discussed in Section I, Chapter 1.

If RPAs truly have changed the nature of war, then we need to re-evaluate our definition of *war*. But a new definition of *war* under this construct would undoubtedly state that RPA crews are "at war" since it was their employment in the first place that was cause for the change in the nature of war. If we assume the nature of war remains enduring and RPAs have only changed the character or means of war, then we need to evaluate whether under the existing definition that RPA crews are indeed "at war." Once again, let's assume RPAs only changed the character of war and take a look at the types of conflict that make up war using the existing definition.

Not Your Granddaddy's War

In the United States, Congress is granted the sole power to declare war. The U.S.'s first declaration of war was made against Great Britain in 1812 and its last declaration of war was against Romania in 1942 during World War II. There have been eleven total declarations of war in U.S. history and none since 1942. So from a semantic (or legalistic) standpoint, despite numerous conflicts in the past seventy-nine years, no one in the U.S. has technically been "at war." From a practical standpoint we know that to be false. So how do we get involved in modern conflicts if they aren't considered wars? Congress passes a resolution that authorizes the use of military force for a specific situation, usually at the request of the president, or the president authorizes the use of military force under, or oftentimes at odds with, the War Powers Act of 1973.

Each conflict since 1942 exists on a spectrum ranging from host nation security assistance, to irregular, unconventional war, all the way up to major theater war. To make matters more confusing, one conflict can also scale up and down the spectrum over time with a changing enemy. As an all-encompassing example, the initial conflict in Iraq in 2003 was a major theater war that turned into a counterinsurgency fight (2004–11), then into a host nation security assistance operation (2011–14), and finally back into a counterinsurgency fight (2014–19). During that period the enemy shifted from the government and military of Iraq under Saddam Hussein, to Al Qaeda in Iraq and other violent insurgent movements, to the Islamic State.

So the nature of the conflicts we have been involved in since the rise of RPAs has been complex, fluid, and dynamic. It's no wonder we struggle with simple questions such as "Are we at war?" To further obfuscate the situation, the RPA community has spent the majority of its time fighting in low-intensity conflicts or irregular wars against non-state actors. These unconventional conflicts look very different from a major theater war in which the enemy wears a uniform and fights with conventional military equipment.

Low-intensity conflicts tend to be fought by U.S. advisers (most often special operations forces) on the ground to support a host nation force, with the majority of their fire support coming in the form of air strikes. In some instances, the conflict involves only air strikes, with no troops on the ground. A 2014 article titled "The US Is Now Involved in 134 Wars or None, Depending on Your Definition of 'War'" described just how perplexing it has become to determine whether we are at war:

> Consider the definition of war put forth by Linda Bilmes (Harvard Kennedy School) and Michael Intriligator (UCLA), who defined war in a 2013 paper as "conflicts where the US is launching extensive military incursions, including drone attacks, but that are not officially 'declared.'" . . . SOCOM [special operations command] admits to having

forces on the ground in 134 countries around the world. That doesn't mean its forces are carrying out capture or kill raids in every country, but it's almost impossible to know where and when different operations are taking place.

By that definition, anywhere there is a drone strike or a special operations team could be considered an area of conflict. In areas where there are drone strikes but no other U.S. forces are present, wouldn't that mean the only combatants "at war" in those conflicts are the RPA crews and the enemy being targeted? In those conflicts, RPA crews most certainly would be considered "at war." And by extension, wouldn't RPA crews also be "at war" in every other conflict in which they are employed doing similar missions?

Retired Marine lieutenant colonel and associate professor at the Naval War College John F. Griffin shared a different perspective with us:

There are four instruments of national power: diplomacy, information, military, and economics. It is through these instruments that nations have the ability to exert their influence globally, support alliances and treaties, and protect national interests. Of these four options, diplomacy, information, and economics are powers that are exercised routinely by departments of the executive branch at the strategic level. It is the application of a nation's military power that should be considered a last resort when all other activities have failed to secure national interests.

Historically, the application of a nation's military strength was considered an act of war. There were few options left to a nation's executive: use the army to invade or use the nation's fleet to blockade or destroy the other nation's fleet. Today, America's military is deployed globally, yet, most of the military actions are not considered acts of war. The positioning and basing of forces in allied nations or at sea often support diplomatic and information efforts, and secure trade and treaties. Since World War II, the employment of special operations forces has been a discreet application of military power that often fell below the threshold of an act of war. The use of these highly trained and specialized small teams to perform clandestine kinetic strikes or build partnership capacity are viewed as a low-risk option (low risk in terms of loss of life, mission failure, and/or provoking a declaration of war by another nation). Today, American authorities at the strategic level have another discreet military option, remotely piloted aircraft.

Remotely piloted aircraft can fly clandestine surveillance missions and even conduct targeted air strikes. Similar to the employment of special operations forces, these actions are generally considered low risk and below the threshold of war. Yet, as a popular option, RPA further reduces the risk of loss of life *or capture* and mission failure. Additionally, the commitment of SOF [special operations forces] can still make civilian authorities nervous due to a lack of control once the option is authorized. Yet, RPA missions provide civilian decision-makers a deeper sense of control right up to the point of execution.

One thing is certain: when we have to debate whether someone who kills an enemy combatant with his weapon system is "at war," that's a sure sign that this isn't your granddaddy's war and most likely won't be anytime in the near future.

The Road to War

I discussed in the previous section the process by which the United States enters into a conflict against an enemy; however, I did not discuss the considerations that lead up to such a decision. Concerning our civilian leadership who make the determination to enter into a conflict, a key question to consider is: "Have RPAs lowered the threshold to enter into war?" Has the perception of a "war without friendly casualties" made us more likely to go to war or is the use of RPAs just a further extension of the trend of using only airpower in certain conflicts to reduce the possibility of friendly casualties? Or are RPAs just another tool in the military arsenal alongside all the other tools necessary to achieve strategic objectives in a conflict? To answer these questions, first I will explore the principles of *jus ad bellum* (Latin for the "right to war") under just war theory.

The formal ideas of *jus ad bellum* (the right to war) and *jus in bello* (the rules of war) gained sanction with the formation of the United Nations post–World War II. But philosophers as early as St. Thomas Aquinas in the late 1200s significantly advocated for a set of rules for war that protected civilians and attempted to ensure states entered into conflict with each other for just reasons. The principles of *jus ad bellum* are considered those just reasons or authorization for why and when a state would enter a conflict. Those principles consist of proper authority and public declaration, just cause or right intention, probability of success, proportionality, and last resort. Of those principles, the most applicable to the RPA discussion are proportionality and last resort.

The principle of proportionality states that only the necessary military force will be used to achieve military or political objectives in war. The goal behind this principle is to limit destruction on a massive scale, ensure only necessary and legitimate military targets are attacked, and assure that caution is taken to protect noncombatants in the process. World War I and (especially) World War II represent the most horrific failures or exceptions to this concept. From a proportionality perspective, RPAs sound like a near ideal weapon for civilian leadership to choose. Bombs and missiles employed by RPAs have relatively low-yield explosives compared to other strike aircraft. Those bombs and missiles are all precision-guided munitions. Due to their endurance, RPAs can watch a target for a long period and strike when the time is favorable, eliminating or mitigating civilian casualties.

Opponents of RPAs, however, attempt to view proportionality in terms of an RPA as the literal replacement of boots on the ground in a country, instead of RPAs supplanting or supplementing other strike aircraft. Their argument is that troops with rifles on the ground in the battle space offer a much lower proportionality "weapon" than an RPA employing a bomb or missile. If that is the correct working definition of proportionality, then taken to its natural conclusion we should employ only troops armed with knives to fight our wars. A soldier armed with a knife is a very precise weapon system that can be employed with almost no possibility of collateral damage. But then again that contradicts the principle of probability of success and, more important, imposes too much unnecessary risk to friendly forces. No, RPAs are an appropriate proportionate weapon, especially when compared to other similar weapon systems: manned strike aircraft.

From a proportionality perspective then, have RPAs lowered the threshold to enter into war? Not really. At face value, a more precise weapon that reduces civilian casualties during attacks, limits destruction to legitimate military targets, and reduces destruction on a large scale would be a preferred weapon that would be more attractive for civilian leadership as a military option.

Fundamentally, there are two issues with this assumption. It assumes that civilian leaders are selecting this specific weapon when making the decision to go to war, instead of making the decision to commit to war and letting military leadership determine the best method to accomplish military objectives. Secondly, it assumes that deciding to go to war and only using RPAs provides for probability of success.

Despite numerous airpower theorists throughout history such as Giulio Douhet, Billy Mitchell, and Curtis LeMay, who collectively thought complete victory in war could be achieved by airpower alone, this has yet to be proven in practice in the wars in which it has been attempted. We may be able to bomb the enemy back to the Stone Age, but that doesn't guarantee victory. In fact, in a counterinsurgency fight, a struggle to win the "hearts and minds" of the population, this method may actually be counterproductive. Without some form of "boots on the ground," massive bombing campaigns often serve to harden the will of the target, as can be seen in the British response to Germany's bombing raids on the UK in WWI and WWII.

In the unconventional wars that RPAs have been employed in over the last two decades, they most certainly have been the only military weapon employed in a few conflicts. And even though RPAs helped to achieve the military objective of killing enemy fighters in large numbers, which as we witnessed in 2019 led to the demise of ISIS, they have yet to win a war by themselves.

Concerning the principle of last resort, there also exists a theory among the anti-drone camp that RPAs lower the threshold to go to war since they embody the perception of a casualty-free war to the side using them. The French philosopher Grégoire Chamayou argues in *Drone Theory* that:

> The dronization of the armed forces, just like any other procedure that externalizes those risks, alters the conditions of decision making in warfare. Because the threshold of recourse to violence is drastically lowered, violence tends to be seen as the default option for foreign policy.

David Cortright in the article "License to Kill" agrees with Chamayou:

> The use of these weapons creates the false impression that war can be fought cheaply and at lower risk. They transform the very meaning of war from an act of national sacrifice and mobilization to a distant, almost unnoticeable process of robotic strikes against a secretive "kill list"...
>
> The possession of drone technology increases the temptation to intervene because it removes the risks associated with putting boots on the ground or bombing indiscriminately from the air...
>
> Any development that makes war appear to be easier and cheaper is deeply troubling. It reduces the political inhibitions against the use of deadly violence. It threatens to weaken the moral presumption against the use of force that is at the heart of the just war doctrine.

Alan Dowd, in "Moral Hazard: Drones and the Risks of Risk-Free War," agrees as well:

> ...drones make it easier to go to war. Thus, war is increasingly becoming a first resort rather than a last resort.... As the risks related to war decrease, it seems the likelihood of waging war increases. "If war becomes unreal to the citizens of modern democracies," political theorist Michael Ignatieff worries, "will they care enough to restrain and control the violence exercised in their name?"... The temptation to gain all the benefits of kinetic military operations with none of the costs, consequences, or risks may be too strong for the Executive to resist.

These are just a few of the opinions expressed that RPAs lower the threshold to enter war. There are numerous others. It is argued that RPAs are viewed by civilian leaders as the perfect solution to war in a risk-averse political environment. The only risk is loss of treasure as there is no skin in the game, no friendly casualties, and therefore no political backlash by the citizens for entering into a conflict.

If this is indeed true, it was airpower in general that gave rise to this mindset of using a low-risk, high-yield military option, not RPAs. The RPA just further embodied it. Operation Allied Force, prior to the existence of armed RPAs, in Kosovo is a perfect example: a NATO-led coalition sent to stop ethnic cleansing solely through the use of airpower. Cruise missiles offer yet another example of a no-risk option. But the trend and desire to seek less risk for our forces in war goes much further back in history. The historian Paul Johnson in *Modern Times* explains that even in World War II strategic bombing "was the best way to make the maximum use of their vast economic resources, while suffering the minimum manpower losses." This mentality and approach are nothing new and certainly not attributable to the rise of RPAs. While airpower can significantly influence the outcome of a war, airpower alone will not win a war. And, yes, RPAs are included in airpower, not separate from it.

While so many critics state that RPAs reduce the threshold to go to war, the truth is there actually is no anecdotal or empirical evidence that RPAs have altered politicians' decisions to enter into conflict or that we have been involved in more wars as a result of RPAs. There are no reports of the commander in chief stating that he made the decision to enter into a conflict because the desired objectives could be achieved without friendly loss of life thanks to RPAs. If this were true, it's highly unlikely that politicians wouldn't use this to a political advantage, and we would hear about it or, more important, have evidence of it.

There also isn't any empirical evidence that RPAs have made the decision to go to war easier. In fact, in all of the conflicts the U.S. has been involved in from 2001 to 2021, in only one have RPAs been employed without troops being on the ground or other strike aircraft being employed in that same country, Pakistan; and even there we have had troops on occasion for limited periods of time. Additionally, you could argue that Pakistani national forces, supporting our battle against terrorists in that nation, represented a form of boots on the ground.

In Afghanistan, Iraq, Somalia, Niger, the Philippines, Syria, Yemen, Libya, and other areas where RPAs were employed against international violent extremist organizations, there were other troops engaged in those conflicts. In these instances, the decision to enter into a conflict did not reduce the risk to other U.S. forces, which is portrayed as the barrier reduced by RPAs, thereby making it easier to go to war.

If RPAs do indeed reduce the threshold to go to war, then why employ other troops and assets as well? Why not just send in the drones? First, because RPAs alone do not provide for great probability of success in any conflict, and, second, because they haven't been employed in that manner. After two decades of continuous fighting around the globe, there are not enough data points to make a conclusion on the assertion that RPAs have reduced the barrier to war. The historical facts just don't back up the conveniently crafted narratives of the anti-drone crowd that RPAs reduce the threshold to enter into a conflict.

Targeted Killing or Assassination?

What about the claim that RPA crews are just targeted assassins, killing specific individuals on a "kill list"? This is where our perception of modern war has not caught up to the changes in the character of war. In past conventional conflicts we targeted military equipment, military units, and military facilities in hopes that when the enemy reached a certain level of combat ineffectiveness or destruction they would either surrender or be completely incapable of waging war. We looked for "centers of gravity" (sources of strength or military might) to destroy in the hope that their destruction would lead to the overall defeat of the enemy. These assumptions are built on a Westphalian state model in which one state fights another while adhering to internationally established agreements for the conduct of war and peace.

But what does that war look like when the enemy is part of an extreme ideology, spread across international borders, funded through criminal activities, and evinces no respect for, recognition of, or adherence to internationally established agreements for the conduct of war? What happens when we are fighting non-state actors such as terrorists, insurgents, and criminals? It obviously changes the method in which that war is fought and the center of gravity that needs to be destroyed.

It is often assumed that the critical vulnerability of insurgencies or terrorist organizations is the leadership. Kill the leadership and the organization flails. That's partly why the methodology in fighting counterinsurgencies over the past two decades has been to kill or capture the leadership and why the RPA community is killing people on a "kill list," giving the perception of being targeted assassins. To be fair, the list is not just for RPAs. It's just a list. Any method that's available and makes the most tactical sense in the situation will work; a manned air strike, a raid to capture, or an RPA strike. And in many cases, high-profile individuals have been captured or killed using other means. Look, for example, at the capture of Saddam Hussein, the manned air strike that took out Abu Musab al-Zarqawi, or the killing of Osama bin Laden by special operations forces.

Opponents to this policy of targeting individuals postulate that these leaders are replaced as fast as they can be killed in these types of organizations and what replaces them may actually be more brutal and violent than the original leader. In short, that targeted killing may work at the tactical level but fails at the strategic level. That is a valid point, but a situation outside the control of those executing at the tactical level, particularly RPA crews. That is a policy-level decision, not a tactical-level decision. In fact, these kinds of situations can lead to a negative experience for RPA crews. Imagine that you spent several hours, days, or weeks watching an enemy leader and finally killed him when the time was right, only to learn that it didn't really matter. He was easily replaced, and not necessarily by someone who we would have wanted to replace him. Your hard work seems in vain, or even counterproductive at that point, and probably gives pause to why the original leader was killed in the first place.

The killing of individual leaders in unconventional warfare is often perceived as a benchmark of progress. It dominates the news cycle for a day. It makes for good headlines when we hear that a "senior level Al Qaeda or ISIS leader" was killed in an air strike. It has become the modern-day equivalent of the enemy body count of past wars. But the truth is we can't kill our way to victory when fighting an ideology. We have to change the conditions to compel the enemy to bend to our will. To be fair, killing is part of the solution to change those conditions, just not the entire solution. It's akin to the problem set that the British Empire faced during the Golden Age of Piracy from the 1650s to the 1720s. The British solved this problem with a combination of violence and policy.

During this period, piracy plagued the world's superpowers, particularly in the Mediterranean Sea and the Caribbean. Piracy, at the time, was a criminal activity just like terrorism in modern times. The British used the Royal Navy to hunt down pirates and enforce policies with local governments, eventually ending piracy on a large scale. So why weren't these criminal activities handled by law enforcement, which tends to have an apprehend approach rather than a kill approach? Why were pirates hunted by the British Navy and why are terrorists hunted by the U.S. military?

The answers are simple: resources, capabilities, opportunity, and budget. The U.S., just like the British Empire in the seventeenth century, is not organized to conduct law enforcement operations on a global scale, but it is arrayed to conduct military operations on a global scale. Just as the British Empire in the late 1600s and early 1700s used its Royal Navy to enforce its policies of anti-piracy, the U.S. is arrayed to conduct global antiterrorism and counterinsurgency operations on a massive scale with its military forces. Even though counterterrorism is traditionally a law enforcement activity, the only forces with the resources and budget to tackle counterterrorism on a global scale is the U.S. and other global powers' militaries, similar to the British Navy during the Golden Age of Piracy.

Another item for consideration is that due to the changing character of war with the employment of RPAs, it is easier to be more discriminant in who we kill. And isn't this actually a good thing? Don't we want to target just those individuals who are enemies of the state, while minimizing collateral damage to those who mean us no harm? Don't we want to be able to kill specifically, and only, those individuals who are our enemies rather than targeting a building, town, facility, geographical area, or unit? Specifically, when fighting an insurgency whose individual leaders may motivate others to take up arms in the cause of their extreme ideology.

The obvious answer is yes. We want to have the option to take the fight directly to those individuals belonging to a non-state actor bent on doing harm to Americans and our coalition partners and allies. We want to be able to target specific individuals, and RPAs offer that opportunity. Just as marketing in the twenty-first century has become personalized and tailored to an individual's shopping and online search preferences, weapons of the twenty-first century have become tailored to personalize the killing of a specific individual. That gives the appearance of targeted assassination, but one should argue that it is more like personalized targeting or targeted killing, with the important distinction of being sanctioned.

The difference between assassination and targeted killing can be summed up fairly simply: those who believe RPA crews are at war and morally, legally, and ethically justified in their actions consider it targeted killing. Those who don't believe RPA crews are at war consider it assassination or murder. I'll give Georgetown law professor Gary Solis the final say with this passage from his book *The Law of Armed Conflict: International Humanitarian Law in War*:

> Assassinations and targeted killings are very different acts.... First, an international or non-international armed conflict must be in progress. Without an ongoing armed conflict, the targeted killing of an individual, whether or not a terrorist with a continuous combat function, would be homicide and a domestic crime. It is armed conflict that raises the combatant's privilege to kill an enemy.

Regardless of which side of the debates mentioned above you fall on, it should be obvious that we are in an unsettled state with the acceptance and understanding of RPAs. A healthy debate over the moral, ethical, and legal use of weapons is definitely a good and necessary thing. Similar outcries in the past led to the outlawing of chemical weapons and land mines and a reduction and regulation of nuclear weapons and ultimately lead to a more just war and a just peace. Unfortunately, across the span of the past two decades, the warriors fighting with this weapon system on a daily basis, navigating this complex environment and public debate, have been the ones most affected and suffering from the unsettled nature of it in the interim.

So are RPA crews at war? Go ask the Reaper crew who blew the legs off an enemy fighter with a missile and then watched his son push him home in a wheelbarrow. Have RPAs changed the nature of war to targeted assassinations? No. RPA crews are at war and therefore from a legal, moral, and ethical standpoint are entirely justified in killing enemy combatants. Assassination is murder, targeted killing is what occurs during a military conflict. RPAs have not changed the nature of war, but they most certainly have changed the character or means of fighting a war. In the next chapter I'll explore how this change in character has impacted our perception of what it means to fight with honor and how it has impacted the enemy's perception of us when we fight with RPAs.

Chapter Two

RPA and the Warrior Ethos

The resentment created by American use of unmanned strikes is much greater than the average American appreciates.
—General Stanley McChrystal, former commander of Joint
Special Operations Command

The real advantage of unmanned aerial systems is that they allow you to project power without projecting vulnerability.
—General David Deptula, USAF

The Warrior Ethos

The President of the United States in the name of The Congress takes pleasure in presenting the MEDAL OF HONOR to

CORPORAL DAKOTA L. MEYER

UNITED STATES MARINE CORPS

For service as set forth in the following:
For conspicuous gallantry and intrepidity at the risk of his life above

and beyond the call of duty while serving with Marine Embedded Training Team 2-8, Regional Corps Advisory Command 3-7, in Kunar Province, Afghanistan, on 8 September 2009.

Corporal Meyer maintained security at a patrol rally point while other members of his team moved on foot with two platoons of Afghan National Army and Border Police into the village of Ganjgal for a pre-dawn meeting with village elders. Moving into the village, the patrol was ambushed by more than fifty enemy fighters firing rocket-propelled grenades, mortars, and machine guns from houses and fortified positions on the slopes above. Hearing over the radio that four U.S. team members were cut off, Corporal Meyer seized the initiative.

With a fellow Marine driving, Corporal Meyer took the exposed gunner's position in a gun-truck as they drove down the steeply terraced terrain in a daring attempt to disrupt the enemy attack and locate the trapped U.S. team. Disregarding intense enemy fire now concentrated on their lone vehicle, Corporal Meyer killed a number of enemy fighters with the mounted machine guns and his rifle, some at near point blank range, as he and his driver made three solo trips into the ambush area.

During the first two trips, he and his driver evacuated two dozen Afghan soldiers, many of whom were wounded. When one machine gun became inoperable, he directed a return to the rally point to switch to another gun-truck for a third trip into the ambush area where his accurate fire directly supported the remaining U.S. personnel and Afghan soldiers fighting their way out of the ambush.

Despite a shrapnel wound to his arm, Corporal Meyer made two more trips into the ambush area in a third gun-truck accompanied by four other Afghan vehicles to recover more wounded Afghan soldiers and search for the missing U.S. team members. Still under heavy enemy fire, he dismounted the vehicle on the fifth trip and moved on foot to locate and recover the bodies of his team members.

Meyer's daring initiative and bold fighting spirit throughout the 6-hour battle significantly disrupted the enemy's attack and inspired the members of the combined force to fight on. His unwavering courage and steadfast devotion to his U.S. and Afghan comrades in the face of almost certain death reflected great credit upon himself and upheld the highest traditions of the Marine Corps and the United States Naval Service.

Our highest military honor in the land is awarded only to those who demonstrate extraordinary courage in the face of great peril. It reflects the value we place on the warrior ethos. From reading Sergeant Dakota Meyer's Medal of Honor citation, it's evident that a warrior ethos still exists today. But does it exist uniformly across the services, ranks, and military occupations? Does a warrior ethos apply to RPA personnel?

Warriors throughout time have been judged, honored, and revered based on their physical prowess, strength, and bravery. Early knights are the standard-bearers for this model with their code of honor and undaunted courage in the face of danger. But every advance in weaponry and every change in the character of war since the age of knights edges us further from the concept of fighting in this honorable way: face-to-face with our enemy, equally invested in the duel with the outcome determined by skill, strength, experience, training, physical ability, courage, and perhaps a little bit of luck. Warriors today may never come face-to-face with the enemy they kill, and advances in the technology of weaponry means that the outcome relies less and less on strength, physical ability, and to some extent physical courage. In fact, as Chamayou discusses in *Drone Warfare*, the outcome may just depend on money:

This [drone warfare] is by no means unprecedented. Every time that, as Voltaire put it, "Whoever was rich became almost invulnerable in war," warfare turned into one-sided killing. As soon as one camp made itself practically untouchable through an overwhelming superiority in weapon, life and death took up their positions in an exclusive fashion on one or other side of the front line.

The sentiment of one-sided killing also existed in the nascent days of submarines, long bows, and sniper rifles. It is common for society to espouse the view that a new weapon has "changed everything" and we have lost the honor in how war is fought. But every weapon follows with either a proliferation of the same or similar weapons or a counter to the weapon leveling the playing field.

So where does this leave us in modern warfare? Is the warrior ethos of today the same as that of warriors who have gone before us? What about someone whose military occupation will never put him in a situation requiring unwavering courage in the face of almost certain death? These are extremely subjective questions. It is like comparing a football player today to players of the helmetless period. Does that simple change in how an act is performed make one group tougher, more revered and honorable in how they executed something at the time? Because we view the helmetless football player as a "tough warrior," does that preclude us from viewing modern-day football players in the same manner since they wear helmets?

Does that also mean that modern warriors in comparison to knights fall short of our perception of a warrior ethos? Or does it just mean that a warrior ethos and our definitions of *tough* and *honor* evolve over time? Do the weapons we employ and how we fight change our perspective on the warrior ethos, particularly as it relates to RPAs? To answer these questions, we need to define a warrior ethos that stands the test of time and technological advancements. Fortunately, Steven Pressfield offers a fairly timeless definition in his aptly named book *The Warrior Ethos*:

> The Warrior Ethos is a code of conduct—a conception of right and wrong, of virtues and vices. No one is born with the Warrior Ethos, though many of its tenets appear naturally in young men and women of all cultures. The Warrior Ethos is taught. On the football field in Topeka, in the mountains of the Hindu Kush, on the lion-infested plains of Kenya and Tanzania. Courage is modeled for the youth by fathers and older brothers, by mentors and elders. It is inculcated, in almost all cultures, by regimen of training and discipline. This discipline frequently culminates in an ordeal of initiation. The Spartan youth receives his shield, the paratrooper is awarded his wings, the Afghan is handed his AK-47.

This code of conduct and conception of what is right and wrong is best described as fighting in an honorable way. Therefore, the warrior ethos is tied to honor. To fully understand the warrior ethos as it relates to RPA personnel we must explore whether RPA crews fight and kill with honor. There are assertions that RPA crews do not fight in an honorable manner—that they are cowards—and that leads to moral hazards and moral injury within the community. This is what the Japanese said about repeating firearms in the mid-1800s, and so they banned gunpowder weapons. That did not go well for them.

To fight and kill in an honorable manner means that the act and the warrior adhered to the legal, ethical, and moral standards established by society. Society regards with great respect those warriors who fight in a manner that it deems honorable. Legally it means that a warrior followed the rules of engagement (ROE) and did not violate any principles of the Law of Armed Conflict (LOAC). Ethically it means the warrior adhered to the accepted code of behavior of society. Morally it means the warrior behaved in a righteous and principled manner. The legal standard is concerned with "can we kill someone?" while the ethical and moral standards are concerned with "should we kill someone?" and "how should we kill them?" Simply put, because we can legally kill someone doesn't always mean we should kill them, nor does it mean we can kill them in any manner we deem necessary. Legal, ethical, and moral standards define the set of preconditions that must be met for us to kill honorably in combat.

Can RPAs Kill with Honor?

To answer this question from a legal perspective, there are several topics to consider. Do we employ RPAs in a manner that violates a state's sovereignty, something that used to be considered an act of war? Does killing individuals with an RPA amount to illegal assassination? What about the legality of killing an American citizen with an RPA strike? What is the legal combatant status of an RPA crew on American soil? And, lastly, as a related subject, what is the legality of killing with autonomous killer robots in the future given we have already come this far? All easy questions, right?

In considering the answers to these questions, first let's establish a baseline for discussion. All modern American warriors—whether remote operators, maintainers, fighter pilots, deckhands, ground pounders, or even finance troops—must adhere to a series of rules to conduct warfare legally. Broadly, these rules are referred to as the Law of Armed Conflict or the International Humanitarian Law. Historically, the law dates back as early as 1864. The Geneva Convention signed the LOAC into international law in 1949, defining a series of principles that outlaw atrocities and the horrific methods of killing soldiers and noncombatants that occurred in WWII, including the sick, the wounded, and POWs. Albeit this is the fundamental core, LOAC is also a compilation of principles from both the Hague Conventions and the Chemical Weapons Convention, as well as the Uniform Code of Military Justice (UCMJ) and the Code of Conduct. Functionally and most germane to our discussion, LOAC sets the principles and standards in conducting war among all members of the United Nations or NATO.

Given this baseline of overarching conduct, and in the particularly litigious and accountable environment that war exists in today, it is no surprise that all modern American warriors would strive to adhere to the legal standards of warfare regardless of location, situation, and weapon system employed. And as proof of this assumption and the efficacy of LOAC, war crimes rarely occur within the modern American military. But when they do occur, regardless of the reason, they are thoroughly investigated and tried. War was not always fought in this manner, with a strict adherence to an internationally recognized set of rules. In fact, these current rules of war exist as a result of past atrocities and methods of killing so horrific they prompted international outcry to prevent such barbarism from reoccurring.

Despite its efficacy, one of the challenges to LOAC is keeping stride with technological innovation in warfare. Historically, legal permissions and restrictions lag behind the development and employment of weapons on the battlefield. RPAs are no exception and the full legal framework to define their employment is still a work in progress.

Returning to our question set, let's address the legality of flying an

RPA within another state's airspace and the assertion that it violates that state's sovereignty. Air Force Intelligence Officer Major Michael Kreuzer touches on this in "Examining the Future of Unmanned Combat Aerial Vehicles and Remotely Piloted Aircraft—Analysis":

> Despite numerous predictions that RPAs could exacerbate conflict by undermining sovereignty and allowing states to violate airspace with impunity (a charge often leveled against the United States for its RPA campaigns), experience to date has largely been the opposite.

Kreuzer goes on to explain that violations of sovereign airspace with RPAs from other nations have been met with stiff resistance, resulting in many being shot down. It is important to point out here that RPAs are extremely vulnerable and relatively easy to shoot down by any nation with a competent air force. This serves as a deterrent for nations that may consider risking their assets to violate the sovereignty of another state.

Despite the appearance that U.S. RPAs are violating sovereign airspace, this is not the case. In most instances, U.S. military forces are employed at the request of the host nation to aid them in a fight against their enemies, whether publicly acknowledged or not. In other instances, there is an enemy that exists within a failed or failing state, incapable of effectively governing, such as Somalia, or within a lawless region of a state, such as the federally administered tribal area (FATA) within Pakistan.

RPA strikes in Pakistan have elicited the most concern and outcry in the media of the U.S. violating that country's sovereignty. During this period, several terrorist attacks against the Pakistani government originated from within the FATA and the Pakistani military conducted operations against the Taliban since 2001. These actions demonstrate the Pakistani government has enemies within the FATA and has turned a blind eye to assistance in eliminating them. This is their wild, wild west under no control or influence from the central government. It isn't difficult to see that there is a lot at play in this area and situation, and it's highly doubtful that America is violating Pakistan's sovereignty with RPA without permission. In fact, if this were the case, Pakistan had ample opportunities over the period of several decades to protest in a manner that would hurt the U.S.'s war effort in Afghanistan, but this has never happened.

So we are able to fly in other countries. What about striking in those countries? In some conflicts, the U.S. has been authorized by its civilian leadership to use force against an enemy such as Gaddafi's forces in Libya or ISIS in Syria. Legally, we don't ask for a state's permission to bomb that state—that's in the fine print of the authorization to use force—but we still abide by LOAC while engaged in that, or any, conflict.

Regarding assassination claims, as addressed in the previous chapter, it is the authorization of the use of force that determines the legality of a conflict, not its means. In other words, it does not matter if lethal force was from a bullet out of a Marine's M4 rifle or a Hellfire missile coming off the rail of an MQ-9, it was authorized and is therefore not assassination. Once a conflict is determined to be legal, killing of a declared enemy is combat, not assassination. Even if that enemy is eliminated during a targeted killing as a result of appearing on a "kill list" and even despite the "ambush" style of the strike. This is most evident in the War on Terror and serves as the legal justification for targeting terrorists belonging to major terrorist organizations such as Al Qaeda, ISIS, etc.

Does the authorization of the use of force give the U.S. the legal authority to engage in the extrajudicial killing of American citizens? It's difficult to see how this is legal in many circumstances. Necessary? Probably. But legal? Definitely subject to debate. In both world wars there were cases of American citizens choosing to fight on the side of our enemy. If U.S. forces had killed these individuals in combat, would that have been an illegal act? Fortunately, this has been reported to have occurred only three times in two decades of global conflicts.

What is an RPA crew's legal combatant status when employing a weapon system from the United States? Could the enemy, if they had the means, legally target them on American soil during the conduct of their mission? What about outside the conduct of their mission? Does an RPA crew lose their legal combatant status when they step outside the box, or when they leave the base and return home? Or are they always considered a lawful combatant, as long as they are in an active RPA squadron at war?

These are issues we have never had to consider with weapons systems employed at the tactical level that unfortunately remain unresolved.

When most warriors are "at war," they are physically present in the combat area and remain in a legal combatant status while in the area, until they return home. That obviously isn't the case with most RPA warriors. Their combatant status may depend on whether they are cognitively "deployed" to combat, meaning whether they are in the box flying a mission. This may become extremely relevant when the U.S. is fighting a peer competitor, but in the current global conflicts against non-state actors, the legal combatant status of an RPA crew may be irrelevant.

And here's why: When fighting non-state actors during irregular wars, our adversaries make no distinction of where the battlefield ends and begins. Given the opportunity to target and kill Americans on our own soil, Al Qaeda, Al-Shabaab, Abu Sayyaf, Boko Haram, ISIS, and several other groups on the Department of State's Foreign Terrorist Organization list wouldn't hesitate to take the opportunity and wouldn't think twice about civilian casualties.

Shane Riza in *Killing Without Heart* claims that this predicament with terrorist organizations exists in part because employing RPAs from our home soil has removed the opportunity for the enemy to kill our soldiers on the battlefield, and it has created a risk inversion that makes our citizens more vulnerable to attack as a result. Riza says that out of a desire to protect our own soldiers we have put our citizens more at risk. It's an interesting theory, but once again one that thankfully hasn't proven true in two decades of employing RPAs in conflicts. Let us also never forget that America was attacked on its home soil on September 11, 2001, well before the rise of RPAs and the possibility of any sort of risk inversion.

One possible remedy to this inversion would be the innovation of autonomous weapons, and maybe this is a viable solution. However, intentionally removing the human from the decision-making loop when it comes to killing is a problem we have yet to face. Make no mistake, we are on the glide slope to one day having lethal autonomous weapons. And based on the pace of technological advancement, that day may be sooner than we think; RPAs are one step closer to bridging that gap.

What remains to be seen is the legality of employing lethal autono-mous weapons. But the legality might be the easiest hurdle to get over considering there is no international law currently regulating it. Just because we can doesn't mean we should, at least not without doing the due diligence to map it out precisely and reasonably. RPAs are a relevant example of this mantra since they are one step closer to that end state of using autonomous weapons. Make no mistake, the difficult ethical and moral challenges relating to the use of lethal autonomous weapons are on the horizon.

Courage and Sacrifice: Are They Even "Remotely" Possible?

If we are abiding by the legal standard when we engage combatants using RPAs, it should be considered an honorable kill. Actually, the legal standard is only part one of the test. The moral and ethical standards must be passed as well, and prove to be more complex than the legal question, albeit there is one rule of thumb: if it is not legal, it's not happening. The legality of whether we can kill someone in combat is binary, a series of "yes" or "no" questions that lead to an answer, and if that answer is yes then it's time to consider the other two components of the test.

Morality isn't that simple. Morality deals with the more subjective question of "should we kill someone?" As an example, imagine an RPA crew is watching the leader of a violent extremist organization in a foreign country. This leader has been responsible for the deaths of many local citizens, including the executions of women and children, and has been responsible for providing funding to fighters who have killed both host-nation and American troops. Assuming that this violent extremist organization has been declared an enemy by the U.S. and that the use of force against the organization is authorized, this leader is a valid legal target. That's the easy part.

Now at what point does the RPA crew kill him from a moral stand-point? Whenever they can find him? When no one else will get harmed in the process, or if only a few people may get harmed in the process?

What if only one other person will die in a strike against this leader, but that person is a noncombatant? Is there a moral obligation to strike this target as soon as they are found because of the danger they pose to host-nation citizens and troops? There aren't necessarily black-and-white, binary answers to the subjective moral questions of "should I kill this person?" Yet these are the decisions our RPA crews are making—or, rather, that are being made for them—as I discuss in the chapter on demands of authority.

These substantial moral decisions are actually where courage and sacrifice come into play for RPA crews. Courage isn't limited strictly to physical courage in the face of danger. Clausewitz in *On War* states that "courage is of two kinds: courage in the face of personal danger, and courage to accept responsibility, either before the tribunal of some outside power or before the court of one's own conscience." So it takes moral courage to answer the difficult moral question "should I kill this person?" Dr. Peter Lee argues in "Remoteness, Risk, and Air Crew Ethos" that RPA operations replace physical courage—the foundation of aircrew ethos since World War I—with moral courage and mental endurance. Jesse Kirkpatrick agrees in "Military Drone Operators Risk a Serious Injury":

> If courage involves engaging in an ethical action that entails a sufficiently high degree of risk—and I think it does—then there is no reason to conclude that drone operators lack courage. If one knows in advance that her job may result in severe moral injury but presses on nonetheless, why would we not consider her courageous?
>
> Drone operators face considerable risk, and they are courageous for doing so. . . .
>
> In the past, new technologies redefined our conceptions of courage; this may very well be the case for the future.

U.S. Air Force RPA pilot Joseph Chapa goes one step further in "Remotely Piloted Aircraft, Risk, and Killing as Sacrifice: The Cost of Remote Warfare" and describes killing remotely for another as a form of sacrifice. Below he describes the difference between someone who kills out of fear of being killed (referred to by Chapa as the proximate killer) and the remote killer who kills on behalf of someone else:

> The self-interested, proximate killer is subjected to personal physical risk, and the psychological effects of killing. The other-interested, remote killer is not subjected to traditional physical risk, but does face a psychological cost. Based on the absence of the kill-or-be-killed forced choice, and the decreased empathetic distance resulting from technological developments, the RPA pilot's psychological risk may be greater than that faced by some ground forces. Much work is left to be done, yet this much is certain: if there is a psychological cost that falls to the RPA pilot while the benefits fall to someone else, then the RPA pilot's action is one of sacrifice.

The expectation of moral injury can potentially cause it to happen. But if killing for others at the risk of one's psychological health is sacrifice and the remote killer takes on that burden, then perhaps it is honorable, despite the distance and the lack of skin in the game or risk to personal injury or death. If an act requires courage and sacrifice on the part of the warrior, it's hard to argue that it isn't honorable and doesn't adhere to Pressfield's definition of a warrior ethos.

That's Not Fair!

From a legal and moral standpoint there is honor in killing remotely. What about from an ethical standpoint? A general definition of ethics is simply the set of moral principles agreed upon by a group of people that governs the specific conduct of behavior of that group's members. Does the tool or method we use to kill an enemy matter from an ethical perspective if we have met the legal and moral obligations? Absolutely. We have an ethical imperative to kill an enemy in a manner that reduces suffering, collateral damage, and unnecessary destruction. RPAs most certainly have the ability to do that. The argument by critics against using

RPAs from an ethical standpoint, however, is twofold: it's unfair to not risk anything in war and it's a coward's weapon.

I once knew a Marine battalion commander in North Carolina who would say "The only thing I know about fair, is that the state brings it to Raleigh once a year." While it's such a simple statement, it reflects a keen understanding that the notion of "fair" does not exist outside of calf-judging contests, carnivals, and cotton candy, and it most certainly does not exist in war.

War is a violent struggle between opposing sides. Removing the opportunity for one side to be equally as violent as the other side has always been an objective of war. It usually means the side with less opportunity to impose its will on the other will most likely lose the struggle. The argument that we somehow owe our enemies the opportunity to kill us in a "fair fight" may adhere to the strict code of a knight's honor, but it is an archaic way of thinking as war hasn't been fought in this manner in hundreds of years. Dr. Peter Lee said it best in *Reaper Force*:

> And this brings me to a curious corner of the public drone debate and a word that regularly crops up in discussion about drones: "fair." As in, "Is it fair to use remotely piloted Reapers against jihadists who can't strike back at them?" Hilarious. The notion of war as a fair fight has emerged somewhere in recent arguments against the use of the Reaper. Since the time of the Chinese military theorist Sun Tzu more than 2,000 years ago—and probably before—the idea has always been to make war as unfair to your enemy as possible. The advantages offered by RPA are not a violation of traditional military strategy—it is what militaries have been after for centuries.

Having personally been on the receiving end of enemy rockets and mortars, I can tell you that my first emotion was anger that the enemy could reach me and I had no available method with which to retaliate. However, never did the notion that it was "unfair" for them to use every weapon they could get their hands on in an attempt to kill me ever cross my mind. War is a violent dance and your dance partner doesn't care if they step on your toes.

Okay, so war is not fair, and RPAs create a definite advantage. What about the proclamation that RPAs are a coward's weapon? Routinely in history we have witnessed those who were themselves unwilling to commit to going into harm's way criticizing those who were for coming up with methods to fight from a safer distance. This is cyclical, as I discussed in the opening chapter on the evolution of killing remotely. Colonel Joe Campo described the process of the introduction of new weaponry as critique, accept, repeat. If someone is still referring to RPAs as cowards' weapons, it's obvious they have not moved beyond the critique stage.

It is highly likely that there are people who will never view RPAs in any manner other than as a coward's weapon. Their notion of a warrior ethos is directly tied to physical courage and strength: traits that are not transparently demonstrated in the RPA community. A combatant who falls outside one's perception of a warrior ethos must necessarily be a coward. In the military, we rarely honor those who overcome a moral challenge as brave, although bravery and courage are exactly what it takes to overcome risk and challenge that can result in bodily harm, including psychological harm. As I will discuss in the chapter on culture, this attitude and perception toward the RPA community, both within and outside the military, has more to do with current culture than anything else.

I started this chapter with the Medal of Honor citation for Sergeant Dakota Meyer. It included a description of intrepidity, audacity, boldness, and physical courage, all in the face of insurmountable odds, fear, danger, and threat to survival. This is what our perception of a traditional warrior ethos is modeled after. But as I have demonstrated, there is room for other warriors to belong to that warrior ethos by adhering to an honorable code of conduct from a legal, moral, and ethical standpoint when fighting. Warriors who fight with the courage to overcome risks to their own psychological harm in defense of others, and warriors who sacrifice in the name of others, adhere to this warrior ethos. These warriors may not fit the traditional mold, but they are worthy of the consideration of honor in their actions.

Chapter Three

Dehumanizing the Enemy Versus Intimacy with the Target

The thing that was so dramatically different than any other mission that I had previously performed was you had an intimacy with the target because of the long dwell and the persistence. In particular on these early missions because we were going after these high-value targets and you would sit on them for a couple of days or more before it was the right time, the right place to pull the trigger.

—Scott Swanson, USAF, First MQ-1 Predator Pilot

Hacking the Mind: Overcoming the Resistance to Kill

Humans have a natural resistance to killing another human being. If you have trouble believing that, read *On Killing*. It has sold more than a half-million copies in eight languages and was on the Marine Corps commandant's required reading list. A central theme of *On Killing* is the human resistance to killing. For armies fighting a war, this resistance can be problematic and counterproductive to a successful outcome. What history has demonstrated is that a series of "hacks" is required to overcome this built-in mental governor. One pertinent example is how soldiers are trained to shoot a rifle at the enemy.

After World War II, the U.S. military changed its rifle range practice targets from a bull's-eye to a human silhouette in the wake of studies showing that the percentage of soldiers who fired their weapon at the enemy was unexpectedly lower than desired. It's a bit ironic that in order to condition humans to overcome their resistance to killing another human targets had to become more lifelike. To counter this human resemblance of the target and further accelerate the conditioning, less time was given to the shooter when firing to consider the target's more human appearance and rifle ranges became more rapid-fire versus slow-fire in nature. The result was a conditioned soldier who, when appropriate and required, could kill another human without hesitation. This is a classic example of operant conditioning at its most basic level.

On the rifle range, a soldier's behavior is positively reinforced through the awarding of points for hitting the target. The highest number of points is awarded for shooting a target in the center mass or the head position, both considered a killing shot. At the end of all the firing strings on the range, an overall score is provided to the shooter. This score may determine how a soldier's career unfolds over the coming year.

For services such as the Army and Marine Corps, an enlisted service member's rifle range score is tied to their overall promotion score. A promotion obviously means more money, more responsibility, and more opportunities; therefore, more points on the rifle range essentially equates to more money in a soldier's pocket. Some services even award soldiers with uniform badges based on their rifle scores, which comes with its own unique set of social ramifications.

As an example, the shooting badge a Marine earns for his rifle range score is displayed for all to see on the uniform he wears when he arrives at a new unit. It is one of the first snap judgments that peers, subordinates, and seniors make of a Marine: how well they can shoot. Justifiably so, since others' lives may depend on this unknown person's ability to shoot a rifle under stress.

There are negative social consequences for earning the lowest shooting badge. Marine Corps shooting badges are categorized as either marksman, sharpshooter, or expert in ascending order of scores. The badges themselves also become more intricate in detail in ascending order. The badge for the lowest rifle score, the marksman badge, is an ugly square box, colloquially referred to as a "pizza box." For shooters who qualify on both the rifle and pistol as a marksman, Marines pejoratively refer to it as shooting "pizza! pizza!," a play on the Little Caesars commercials and a quick way to identify someone who can't shoot very well. In a profession of arms, this isn't how someone wants to be identified, so it reinforces the institutional and social pressure to shoot well.

The rifle range, the most basic form of training service members to kill the enemy with a firearm, is just one example of how the military has conditioned its soldiers to overcome their natural resistance to killing the enemy. The rifle range includes reinforcing rewards for success and visible social symbols of performance. Both methods bolster the conditioning required to overcome the resistance to killing. There is, however, more to overcoming a resistance to kill than just the shape of a target, how much time there is to shoot the target, or what shooting badge is worn on a uniform. More paramount is how warriors are conditioned to think about the enemy.

Oh, the Humanity of It All!

The more humanity that is observed in the enemy, the easier it is to relate to them as a person, and the more difficult it becomes to kill them. For this very reason, in order for warriors to effectively do their job, which may involve killing, they must deliberately dehumanize their enemies in several different ways: through emotional distances, cognitive distances, social distances, moral distances, cultural distances, physical distances, and mechanical distances. Each of these distances factors into an empathetic distance, a measure of the ability to dehumanize the target. Retired Marine Colonel G. I. Wilson described the essence of dehumanization as a method to overcome resistance to killing one's enemy in "The Psychology of Killer Drones—action against our foes; reaction affecting us":

> Dehumanization involves obscuring and/or distorting the human identity and qualities of an enemy (or victim) that are either known or unknown to the perpetrator of violence. The enemy (or victim) is seen as nothing more than objects—rather than anything human...
>
> Thus, dehumanization protocols may facilitate violent behaviour....
>
> Dehumanization can aptly be likened to a two-way street where it is irrelevant whether the "target" is known or unknown just as long as it is dehumanized.

To place empathetic distance between the warrior and the enemy, the warrior must think of the enemy as less than human. Two primary distance factors come into play here: cognitive distance and emotional distance. If you can conceive of the enemy as different from yourself and can no longer perceive the same emotions and feelings as the target, dehumanization is achieved. One of the major ways this dehumanizing process occurs is by changing what the enemy is called into something particularly degrading or demeaning.

History is replete with examples of demeaning nicknames given to an enemy as a method of dehumanizing him. In World War I the British referred to the Germans as Fritz, Jerry, boche, and Hun. In World War II, the Allies referred to the Germans as krauts and the Japanese as nips or slants. But the Allied forces weren't the only ones making disparaging remarks about their enemies in an attempt to dehumanize them. The list of American nicknames given by the Germans and Japanese included Joes, foreign devils, dogfaces, yanks, round eyes, and Amis (short for Americans).

During the Vietnam War, North Vietnamese were referred to as Charlie or gooks, and during the recent conflicts in Iraq and Afghanistan the enemy was often referred to as a haji. To a Muslim, a haji is a term of respect and describes someone who has made the pilgrimage to Mecca, one of the pillars of the Islamic faith. In Iraq or Afghanistan, however, U.S. service members misappropriated the term *haji* to demean an entire population of people by marginalizing a central focus of their identity—their religion.

Regardless of the practical use and historical prevalence in combat, this kind of vernacular falls outside the bounds of acceptable behavior in society today. War, however, occurs in a space where the normal rules of society are suspended. War exists in an environment where worlds and cultures collide in a violent struggle for primacy. That struggle involves killing, which is partly enabled by dehumanizing the enemy by stripping away their identity and replacing it with a demeaning, less-than-human epithet.

These sobriquets for the enemy don't always have to be demeaning in nature to deny humanity, though. In fact, due to the professionalism instilled in RPA aircrews, it is uncommon practice to hear a pilot or sensor operator refer to the enemy using a derogatory name during a mission—for a couple of reasons. First, it is unprofessional, and with voice communications of the crew constantly being recorded these negative monikers don't frequently seep into conversation without being squashed by leadership. More important, though, it's really unnecessary to use a nickname to dehumanize a person being targeted with an RPA.

The work to dehumanize the enemy has already been done through the extensive use of military jargon and brevity codes. As an example, just to name just a few of those descriptors: a MAM (military aged male), a squirter (someone fleeing an objective area after it has been struck with a missile or bomb), or an objective name assigned to an HVI such as Objective MANHATTEN, or simply "target." These terms are so far removed from a normal human description that it is unnecessary to conjure up a demeaning term for the enemy to mask their humanity; instead, the enemy is objectified by these words.

Another way in which war is made more palatable is by using neutral, innocuous words to mask violent deeds. Rarely are the words *kill* and *person* used in tactical communications when striking a target. Unless a JTAC is being shot at and under intense stress, it is not common practice for them to direct the RPA crew to "kill that person standing next to the building." Instead, the verbiage is abbreviated and sterilized to "You are cleared hot to engage the target fifteen meters east of the building." Short, concise, sterile, yet dehumanizing all the same.

There is a pattern to the conversation during requests for an air strike. Each party involved understands the flow of the verbiage and what should come next. So simple terms are used for efficiency and commonality of expression in these communications, and the dehumanizing aspect has a collateral effect of making it easier to kill the enemy.

Seldom do warriors use the term *kill* in training, planning, execution of a mission, or during a debrief of the mission. It was very rare during the interviews I conducted for RPA aircrew to use the word *kill* when describing their engagements with the enemy. Instead, *kill* was replaced with commonly used military jargon such as *strike, engage, remove from the battlefield, take them out,* and *shwack.* These are just a few of the many veiled and watered-down descriptors used for the act of killing.

So, although the act is the same and the outcome is the same, to "strike a target" has a different implication than to "kill a person." The military terms are clinical, sanitized descriptions that enable the warriors, at the critical moment when lethal action is required, not to overthink what they are about to do. It is yet another method of overcoming a human's natural resistance to killing. Replacing *kill* and *person* with other quasi-synonymous words is an effective way to create emotional and cognitive distance from the enemy.

The Moral High Ground

The importance of moral distance in war can't be overstated. War tends to be oversimplified into a two-sided event: a fight of right versus wrong, good versus evil. This oversimplification derives from *jus ad bellum* or just war theory. It is assumed that if both the elected civilian leadership of the government and the military leadership do their jobs correctly, conflict will never be entered into unless it adheres to the principles of *jus ad bellum*: proper authority and public declaration, just cause and right intention, probability of success, proportionality, and only as a last resort.

It follows that if a warrior is already engaged in a conflict, his assumption is that it is a just war, with particular emphasis on the principles of just cause and right intention. When a warrior fights for a cause that is just, what does that make their enemy's cause? When a warrior fights for the right intention, what does that say about the intentions of their enemy? The warrior's assumption is that the enemy's cause is unjust and their intentions are evil, therefore the enemy themself is evil. Killing someone viewed as evil is easier to rationalize; in fact, it elevates the killer to an almost hero-like status in their mind.

Considering the enemy evil is yet another label that serves to veil the humanity of the enemy and enable the dirty business required to fight wars, to kill for a just cause. This concept is why the top four responses to killing among the RPA aircrew I surveyed were a sense of mission accomplishment, feeling like they were just doing their job, satisfaction, and pride. They were fighting for a just cause.

There is a line of thought, however, that RPA warfare is more akin to moral disengagement than simple moral distancing. In *Social Beings: Core Motives in Social Psychology*, Susan Fiske describes moral disengagement as separating moral reactions from inhumane conduct and disabling the mechanism of self-condemnation. Detailing this assertion, the psychologist Albert Bandura in *Disengaging Morality from Robotic Warfare* argues that drone crews experience moral disengagement at the behavior locus, the agency locus, the outcome locus, and the victim locus (locus here represents the center or place where something occurs):

> At the behaviour locus, people sanctify harmful means by investing them with worthy social and moral purposes. Righteous ends are used to justify harmful means. Harmful conduct is also rendered benign or even altruistic through advantageous comparison. Belief that one's harmful actions will prevent more human suffering than they cause makes the behaviour look altruistic. Euphemistic language in its sanitising and convoluted forms cloaks harmful behaviour in innocuous language and removes humanity from it. These three mechanisms, operating at the behaviour locus, are especially powerful because they serve a dual function. They

engage morality in the harmful mission but disengage morality in its execution.

At the agency locus, people evade personal accountability for harmful conduct by displacing responsibility to others and by dispersing it widely so that no one bears responsibility. This absolves them of blame for the harm they cause.

At the outcome locus, perpetrators disregard, minimize, distort, or even dispute the injurious effects of their actions. As long as harmful effects are out of sight and out of mind there is no moral issue to contend with because no perceived harm has been done. At the victim locus, perpetrators exclude those they maltreat from their category of humanity by divesting them of human qualities or attributing animalistic or demonic qualities to them. Rendering their victims subhuman weakens moral qualms over treating them harshly. Additional moral disengagement at the victim locus involves blaming the victims for bringing maltreatment on themselves or attributing it to compelling circumstances. In this mode of self-exoneration, perpetrators view themselves as victims forced to behave injuriously by wrongdoers' offensive behaviour or by force of circumstances.

Bandura theorizes that this moral disengagement for RPA crews is distinct from a stress disorder and that continuous self-critique of their actions is what leads to "moral injury." At the same time, Bandura admits that the RPA community is very secretive and that he doesn't actually know RPA crews' attitudes, values, or how they manage their intrapsychic life. The information about RPA aircrew that he bases most of his theories on is taken from biased anti-drone sources, which tend to dominate the media.

Bandura asserts, based on his secondhand knowledge of RPA crews, that they switch their moral control on and off on a daily basis when they go to work. But based on my interviews with RPA personnel, it would be a stretch to conclude that they are morally disengaged. In fact, quite the opposite. It's not moral disengagement that may be causing moral injury for RPA pilots, it's actually moral engagement: understanding the implications of killing, taking responsibility for their actions, and observing the humanity of their victims.

Cultural and Social Distances: You're Not Like Us

Cultural and social distances between warring factions are another way the enemy is dehumanized. Differences in religion, language, race or ethnicity, values, beliefs, and social norms all serve to move the enemy further away from one's perspective of what it means to be human. It can be challenging to relate on a personal level to someone who speaks a different language, has different beliefs, worships differently, and behaves in a manner that is foreign to your perception of what is normal. Essentially, each difference further distances the enemy from our concept of what normal human behavior is, which in turn dehumanizes the enemy and makes it easier to kill him. "They're not like us and their cause is unjust."

Physical Distance: Almost Irrelevant for RPAs

Physical distance plays an important role in a warrior's ability to kill the enemy. This was discussed in great detail in *On Killing* and will be only summarily discussed here. In short, the farther the physical distance from the enemy, the less resistance to killing the enemy one experiences. That observation still rings true for the majority of warriors today, with one major exception. At the time that observation was initially recorded, RPA warfare did not exist to the extent and capability that it does twenty-four years later. Technology has enabled a unique situation never before experienced in the history of warfare, extremely remote killing at the tactical level.

As we have seen with RPA crews, whether seven, seventy, or seven thousand miles away from the enemy, the physical distance to the enemy is not relevant to their resistance to killing and the physical distance might as well be infinite. It is the RPA aircrew's empathetic distance to the enemy that matters in determining the resistance to killing. This comparison between physical and empathetic distance to the enemy will be discussed in detail in the next chapter.

Mechanical Distance: No Longer a Buffer

The last way in which to overcome the resistance to killing the enemy is through mechanical distance using an optic device such as a sight, scope, or screen. Situations viewed live and unaided with an optical device naturally elicit the most emotional responses, whether watching a sporting event, a concert, an accident, or the killing of an enemy soldier. Whenever we can place a barrier between ourselves and the live incident, the situation starts to feel a bit more surreal. It used to be true that night sights for weapons almost entirely masked the humanity of the target due to low resolution of the sensor that made human features less distinguishable. But technological advancements have improved night sights to the point where that is no longer the case.

Early video monitors for RPAs were similar to early night sights with grainy feeds that often cut out and poor camera resolution that made it difficult to discern recognizable features of a human. Targets appeared as blobs rather than people. This unfortunately is where society thinks RPA warfare is stuck at today, pushing a button from seven thousand miles away and firing a missile at a blob on a screen until it's extinguished like in *Space Invaders*. While that may have been closer to the truth in the very early days of using RPAs in the military, we are a few generations past that point today.

Technology created this mechanical distance but, through advancement, technology is now shrinking the mechanical distance. What would have been a buffer between killer and victim is now oftentimes replaced with high-definition video displaying details of the victim never before seen by a warrior during a lethal engagement unless he was fighting hand to hand. High-definition cameras piping in zoomed-in video and pictures on high-resolution screens are actually having the opposite effect of distancing the warrior from the enemy. They are creating a situation where the warrior is becoming more intimate with the target before killing him and observing vivid details in the aftermath of killing him.

An Unexpected Intimacy with the Target

What we are seeing in modern-day RPAs is a struggle between the need to dehumanize the target for the psychological health of the striker and the development of an intimacy with the target prior to and after striking it. To fully understand the situation that RPA aircrew experience, we need to have an understanding of how and why this intimacy with the target develops and what the ramifications are.

In theory, the physical distance of RPA crews from the enemy should be sufficient for them to overcome any resistance to killing, much as was true with other crew-served weapon platforms. But something else is at play here that doesn't involve physical distance, a relationship that is formed between the predator and the unknowing prey. How and why does this relationship form and why is it important? From *On Killing*:

> When you have cause to identify with your victim (that is, you see him participate in some act that emphasizes his humanity, such as urinating, eating, or smoking) it is much harder to kill him, and there is less satisfaction associated with the kill, even if the victim represents a direct threat to you and your comrades at the time you kill him.

While it is obvious that no victim would represent a direct threat to an RPA crew, what is pertinent is the humanity that RPA crews observe in their victims prior to a strike. There has never been an airborne platform in the history of warfare that allows the crew to observe its targets for such a long time with such vivid detail before striking. It is unprecedented and we should expect unprecedented issues as a result.

The relationship between the RPA crew and the target forms as a result of three unique capabilities of RPAs: long flight endurance, fidelity of the cameras, and the ability to schedule the crew during the same shift on a routine basis. The long endurance of RPAs enables them to perform certain missions much more effectively than a manned platform could; this includes pattern of life (POL) observation during an intelligence, surveillance, and reconnaissance (ISR) mission.

While conducting POL, the crew watches a slice of the population in an area to determine a normal baseline of the population's behavior and activities. This helps establish the context necessary to determine when something abnormal happens in that area. These areas of surveillance are usually carefully chosen based on locations of known or suspected enemy activity. It sounds a bit voyeuristic and it is.

Intimacy with the target is determined by knowledge of the target coupled with monitoring the target over time. The greater the knowledge and the longer the observation, the more intimate the connection to the target. Why does it happen? A sense of agency. A recognition of humanity. This is why we identify with characters in a television show that we routinely watch and why RPA crews that routinely watch individuals may form a one-way emotional connection to them.

According to Keith Hillman in "Why Do We Identify with Fictional Characters":

Actually, all it takes for us to "care" about a character is a sense of "agency." This comes from our evolutionary tendency to (mistakenly) project our own thoughts, motives, and emotions onto other people and things...This then extends to empathy—which is controlled by "mirror neurons" in our brain. When we see someone happy, upset, or victorious, neurons fire in our brain in accordance with that and we *feel* some of those emotions. This is all part of an evolutionary mechanism that is simply in place to help us work as part of a wider community and to get along with others (which is crucial to our survival). This works best if said character has a face—as that's where we denote most emotion from (though an "estimation" of a face works well too).

Conducting POL can lead to a controversial "signature strike," a strike that occurs based off the signature or pattern of behavior exhibited by an individual that, when coupled with other sources of intelligence, creates an extremely high probability that this is a legitimate enemy fighter. By no means is this an infallible system for targeting an enemy, but nothing in combat is, especially with non-state actors who blend into the populace, don't wear a uniform, and don't adhere to the Geneva Convention.

Despite how signature strikes are portrayed in the media, they are not indiscriminate killing. In fact, the long endurance of RPAs enables tactical patience before striking a target and great care is taken to evaluate all aspects of the situation before the decision is made to strike the target.

What affects the psychology of RPA crews most about signature strikes is the argument that signature strikes are indiscriminate, illegal, and immoral killings and through association that RPA crews' actions must also be wrong, that it isn't a "good kill." This is one of those areas where RPA warriors are told by society to feel bad about what they are doing, yet are required to keep doing it as part of their profession. It has the potential to create a moral struggle that may lead to moral injury.

The intimacy formed with the target makes it hard to deny the humanity of the target. Whether that intimacy comes from being physically close enough to see the whites of their eyes, or emotionally and cognitively close enough to know their children's names, closeness to a target makes it more difficult to kill. And no mission evokes more connection to a target for an RPA crew than hunting a high-value individual (HVI).

Hunting High-Value Individuals: The Most Intimate RPA Mission

HVIs are often tracked for days, weeks, months, or even years before being struck. This provides ample opportunities to get to know a target through observation. Aaron Garman, a U.S. Army Gray Eagle pilot and sensor operator described the nature of this intimacy with the target in an interview:

> It's ridiculous the idea that we don't see the humanity. I'm watching a target for eight hours. I'm going to watch him go to the store and go

to his wife. I've had targets that I followed for four or five days. I know where they live, I know what they do. I know what they do to the point that I know when we are going to shoot them because that's the time he drives that one road back from the market to his home.

It got to a point where we would make up names for these guys. We would start having conversations for these guys like "Okay, honey, I'm going to leave and go do some ISIS shit now." And she [his wife] would say "Okay, bye, see you at dinnertime." Then eventually you kill this guy. Absolutely I know that his wife is out there and that we just made her a widow and that we just took a father away from his three kids. It sucks. I wish it was just a guy in a car that we didn't know. Those shots are easy.

The nature of shift work often means that crews are observing the same person on a routine basis. Imagine spending eight to ten hours a day watching every move a person makes, every place they go, every interaction they have with the people around them including their family, and every moment where their humanity is on full display. It would be near impossible not to feel an empathetic connection to that target. Again, from *On Killing*:

> At close range the resistance to killing an opponent is tremendous. When one looks an opponent in the eye and knows that he is young or old, scared or angry, it is not possible to deny that the individual about to be killed is much like oneself.

Close range in the instance of the RPA crew is therefore close empathetic range versus physical range. Joseph Chapa, a USAF RPA pilot, explained in "Remotely Piloted Aircraft, Risk, and Killing as Sacrifice: The Cost of Remote Warfare" that "RPA are deconstructing traditional distance relationships. As these weapons push their pilots to the greatest physical distances possible, empathetically, they pull the pilots in, closer than pilots have ever been." The authors of "Psychological Dimensions of Drone Warfare" postulated that "operators may exist in a psychologically dissonant state where there is disconnection and removal from the battle-ground but simultaneous feelings of proximity and intimate connection with targets' lives."

Blair and House in "Avengers in Wrath" refer to this intimacy with the target as Cognitive Combat Intimacy (CCI) and describe it in a brilliant manner:

For remote warriors, *Cognitive Combat Intimacy (CCI) is a relational attachment to a human target mediated by sensor resolution and dwell time, or duration of observation.* In layman's terms, resolution is the clarity with which a hunter can see a target, and even from a great distance, the hunter is exposed to very human factors such as the color of their clothing, the target's interaction with others, and the activities of their daily life. Dwell time can be hours, days, and often weeks, during which the hunter has the opportunity to further develop the empathetic bond with the target. In the case of quick-reaction CAS, dwell might be as simple as watching enemy soldiers hunker down and fire at friendlies. It is the intensity of the action and the relationship to comrades, rather than the duration, that leaves an imprint. Resolution presents the target to the brain as a person, and dwell time allows that knowledge to soak into the consciousness.

This connection to the target is present and it is real. It is similar to the empathy that a hunter feels for his prey after stalking it, waiting for the precise moment to shoot, and killing it. Anyone who has ever hunted knows that there is a mixed bag of emotions associated with a kill. Brad Smith described those feelings of a hunter immediately after a kill in the article "Hey Hunters, It's Okay to Feel Sad After Making the Kill":

If a picture could be taken of the hunter seconds after their killing shot was fired, a whole different aura of hunting would be on display. Remorse, guilt, and sadness would be captured in every image...It is okay for a hunter to feel sad after killing an animal. We are all only human. However, for a man to hunt and be involved in the act of killing an animal and not feel at least some remorse, they should put down their gun and never hunt again.

The hunter brings up an excellent point for hunters of humans as well. There should be emotions associated with killing another human being, regardless of how evil that person is, how much we think they deserve to die, and how many people they have hurt. We never want our soldiers to lose touch with their humanity or the humanity of those they fight. We want disciplined warriors able to deal out controlled violence when required, not psychopaths, fighting our wars.

This is the fear of RPA warfare: that emotionally detached, physically safe drone warriors are killing without remorse, attachment, or compunction. My observations and discussions with RPA warriors contradict this fear. Taking someone's life is an emotional flood, regardless of the distance from which it occurs. That flood of emotions doesn't mean killing shouldn't occur, it simply means there may be an emotional cost experienced by the predator when it does occur.

This is consistent with what I was told during the interviews I conducted. One intelligence analyst I interviewed previously worked for the CIA's RPA program and chose to remain anonymous. As an intelligence analyst for the agency he worked almost exclusively on high-visibility missions hunting top-tier HVIs. He told of two opposite ends of the emotional spectrum that he felt as a result of killing HVIs with RPAs.

In the first case, he and his team had been developing a top Al Qaeda leader for five years when they finally caught a break one day. The analyst said there were about one hundred people in the room watching the mission unfold in real time. At some point during the day, that Al Qaeda leader got himself in the wrong isolated place, which was exactly the opportunity needed to conduct an RPA strike. The room, normally full of quiet professionals, erupted in cheers. It was an emotional day five years in the making.

The analyst's second emotionally charged day at the opposite end of the spectrum concerned another high-value individual that his team watched for six months, for twenty-four hours a day. Every day, the analyst and his team watched this guy walk his kids to school and then go to meetings with other nefarious characters. In the afternoon, they observed the HVI pick up his children from school and then spend hours playing with them in the backyard. According to the analyst, who was a father himself, "There was no doubt that he was a good father."

The analyst knew everything about the HVI and his family. He had pictures of their faces plastered on his workstation. This was the all-encompassing environment that the analyst lived in for six months, just watching and waiting for the right opportunity to kill this man, every day watching him do his dirty business bookended by normal human activities that any father in America might engage in. The analyst said that when the time came to strike this guy, whom he had observed being a normal dad every day for six months, it was emotionally difficult.

As one USAF sensor operator succinctly put it in an interview, "It's the humanity aspect that makes it hard. To overcome that feeling of killing a normal guy you need lots of information about the bad things he does to help justify this killing in your mind." If the target is a high-value individual, it's a certainty that he isn't a boy scout.

Brett Velicovich was an intelligence analyst for the Army Special Forces who used RPAs to help hunt high-value individuals throughout Iraq, including some top-tier targets. He wrote the book *Drone Warrior* to describe his operations. In an interview with *Vox* he described how it felt to watch bad people do normal things:

> You're watching these guys and they're totally normal. You see them dropping their kids off at school. You see them having tea or coffee at a local market. You see them doing normal things. It's almost like *People* magazine or something. You always have these "the stars are just like us" type of feelings. You see terrorists doing stuff that anyone else would do. It's what they're doing in the shadows that we're trying to find. When you find that, then you know you've got him.

Hey, You Killed My Target!

Another strange phenomenon that occurs from such a one-sided intimate relationship between predator and prey is a feeling of ownership or entitlement to kill that individual. This is perhaps the strangest form of intimacy described to us by many RPA aircrew. One USAF sensor operator told of a high-value individual he had hunted eight hours a day for nearly six months. As a result, the sensor operator knew his target extremely well. In fact, he said he could pick this individual out of a crowd solely based on his mannerisms and hand gestures.

At around the six-month mark of hunting this HVI, the sensor operator took two routine days off. Upon his return to work, he was informed that his HVI was killed by another crew during his absence. He described his reaction as a mixture of sadness and anger. While we might think that the sensor operator would be happy that an important enemy figure was killed, his first reaction was a sense of loss, as though someone had stolen something from him. As he put it, the loss he felt was not for the individual, but rather because he didn't get to kill the HVI after investing so much of his life building up to the moment when he could.

Imagine you worked very hard on a project only to have someone come along and complete the last 1 percent and get the credit for all your hard work. That's the nature of shift work when watching an HVI twenty-four hours a day. Someone is going to be disappointed that he didn't get to kill his target.

RPA warfare is a constant balancing act between dehumanizing the enemy and observing the enemy's humanity. Such a struggle has never existed to this extreme for a warrior, particularly at this distance from the enemy. While it's evident that attempts are made to dehumanize an individual, when you watch a person interact with his family it's almost impossible to deny their humanity. That doesn't mean that the acts they have committed are forgivable or that we should let them go because they have a family. It simply means that RPA warriors often find themselves killing someone they can empathize with and who has the potential to lead to negative responses to killing both short-term and long-term in some cases.

Chapter Four

Distance from the Target

Unless he is caught up in murderous ecstasy, destroying is easier when done from a little remove. With every foot of distance there is a corresponding decrease in reality. Imagination flags and fails altogether when distances become too great. So it is that much of the mindless cruelty of recent wars has been perpetrated by warriors at a distance, who could not guess what havoc their powerful weapons were occasioning.

—J. Glenn Gray, *The Warriors*

Physical Distance from the Target and RPAs

Distance between killer and victim plays an immensely important function in killing. In the past, as Glenn Gray described in *The Warriors*, conventional wisdom viewed killing as a direct correlation between physical distance and cognitive distance. The farther the physical distance from the victim the less resistance or hesitation to killing. In *Acts of War*, Richard Holmes described this relationship: "The act of killing is often so blurred by the distance separating killer and victim that it seems like a game or is swamped by a feeling of technical satisfaction in marksmanship." As discussed in *On Killing*:

> The link between distance and ease of aggression is not a new discovery. It has long been understood that there is a direct relationship between the empathic and physical proximity of the victim, and the resultant difficulty and trauma of the kill. This concept has fascinated and concerned soldiers, philosophers, anthropologists, and psychologists alike.

Carl von Clausewitz also eloquently described this relationship between emotions associated with the act of killing and physical distance from the target in *On War*:

> Weapons with which the enemy can be attacked while he is at a distance are more instruments for the understanding; they allow the feelings, the "instinct for fighting" properly called, to remain almost at rest, and this so much the more according as the range of their effects is greater. With a sling we can imagine to ourselves a certain degree of anger accompanying the throw, there is less of this feeling in discharging a musket, and still less in firing a cannon shot.

Figure 3 on the next page from *On Killing* depicts this relationship between resistance to killing as a function of physical distance from the target. At the far left of the spectrum, described as the sexual range, killing is personal, messy, emotional, and undeniably violent. The perpetrator is close enough to the victim to hear his screams, sense his emotions, smell his fear, feel his body go limp, and see the life fade from his eyes. It is extremely intimate and comparable to the closeness and intimacy experienced during sex. It takes a special warrior to overcome resistance to killing at this range. Killing at the sexual range is not for the faint of heart.

At the opposite end of the spectrum, at max range, is a long-range bomber or artillery crew. At max range, the killer may never see his victim, won't hear his screams, and won't have much, if any, of a cognitive or empathetic connection to the victim. In the case of artillery, the crew is firing at targets reported to be at a certain location, normally well beyond visual range. For bombers, bombs are dropped on GPS coordinates or guided to a target via a laser. Modern bombers have targeting pods that do permit observation of the target before a strike and the carnage post-strike, but that time spent observing a target before and after is fleeting due to limited endurance and combat radius of the manned aircraft.

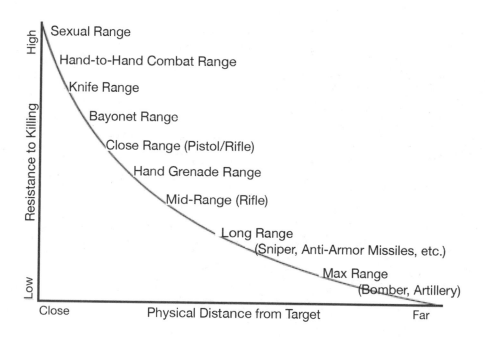

Figure 3

At some point on this spectrum, physical distance becomes irrelevant to resistance to killing. That may occur around the max range and beyond out to what I will call extreme max range, where killing with armed RPAs occurs. Physical proximity to the target is irrelevant from max range to extreme max range. RPA pilots who have also flown fighters and bombers have confirmed this in interviews: at a certain point, distance from the target just doesn't matter.

Arguing about degrees of remoteness beyond the point of maximum distance from the target is a futile endeavor. Once the warrior has determined that he is remote and relatively safe from danger, the proximity to the target no longer matters—whether the target is thirty thousand feet below or seven thousand miles away. Updating Figure 3 based solely on physical distance from the target including RPAs, we see the positioning of the RPA on the chart in Figure 4 varies based on line-of-sight or beyond-line-of-sight capabilities. But all RPAs fall at the long-range or greater category on the line, with armed RPAs at the extreme maximum range portion of the spectrum.

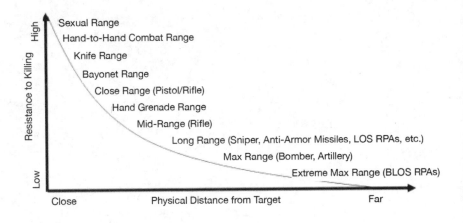

Figure 4

It would therefore follow that, solely based on physical distance, there should be little to no resistance to killing with an RPA according to Figure 4. Of the 243 warriors I surveyed who had used an RPA to kill, 94 percent of them said they thought that killing with RPA might be easier than killing in person. Although most of my respondents lacked a frame of reference for killing other than with an RPA, their perception squares with Figure 4. RPA crew members with previous experience flying fighters or bombers didn't perceive a difference between varying degrees of distance from the enemy during the kill. But physical distance is only part of the equation that needs to be considered when we look at how RPA crews respond to killing.

Mistaking Physical Distance and Safety for Disengagement

The physical distance between RPA crews and their targets has led some critics to postulate that at extreme maximum range killing is an insignificant, inconsequential, and a trivial act akin to pushing a button while playing a video game. Philosophy professor Laurie Calhoun subscribed to this line of thought in "The End of Military Virtue":

> Moreover, there is no empirical distinction to the killer himself between what he does in obliterating a target and what he does in playing a video game...The emotions associated with the activity of killing and risking death have been progressively muted with distance and now eliminated from the act altogether in summary executions effected by UAVs and managed by desktop warriors....Killing from vast distances with the click of a computer mouse, an action so trivial and perfunctory that it is used also to send e-mail and shop online, can only have the effect of altogether insulating killers from the reality of what they do.

Calhoun further describes RPA warriors as assassins who can't recognize the sufferings of their victims because the killing doesn't seem real from their perspective. She views RPAs at the far right of the spectrum in Figure 4: extreme maximum physical distance with no resistance to killing or physical or psychological impact on the killer. Not only did Calhoun get even the basics completely wrong—armed RPAs use a control stick with a trigger for weapons release, not a computer mouse—but, more important, what she fails to recognize is that physical distance between victim and killer is only one aspect of the distance that needs to be considered.

Another popular opinion, expressed by the psychiatrist Theodore Nadelson in *Trained to Kill*, is that "Technology removes the soldier from personal involvement. It removes the passion from killing and the soldier may feel less guilt responsibility." But with RPAs it is technology that simultaneously increases the physical distance between killer and victim and shrinks the empathetic distance and mechanical distance. My research has indicated that the models in Figure 3 and Figure 4 do not represent what is happening with an RPA crew, and it is important to remind ourselves that Figure 3 was developed before this technology and method of killing even existed. Therefore, we must consider a new model that accounts for the nuances of remote warfare.

Rethinking Distance and the Impact on the Killer

Why do RPAs break the paradigm of Figures 3 and 4? U.S. Air Force RPA pilot Joseph Chapa suggests in "Remotely Piloted Aircraft, Risk, and Killing as Sacrifice" that "technological developments have shattered the one-to-one model." Technology that did not exist when Figure 3 was developed has led to a situation where killing occurs at maximum physical distance and extremely close empathetic distance. Physical distance removes the killer from danger, but cognitive distance (synonymous with empathetic distance) brings the reality closer to home.

It is frequently said within the RPA community that an RPA crew's distance to combat is not seven thousand miles but rather eighteen inches, the distance between the crew and war unfolding on the screen in front of them. The French philosopher Grégoire Chamayou in *The Theory of the Drone* theorizes that "this novel combination of physical distance and visual proximity gives the lie to the old [Clausewitz-Hegelian] law of distance...increased distance no longer makes violence more abstract or more impersonal but, on the contrary, more graphic and more personal."

Technology has enabled twenty-first-century RPA warriors to kill an enemy combatant, see the horror of his body being blown apart or his blood spewing everywhere, watch his heat signature escape from his body as he dies, and watch those that come to mourn his death—all in zoomed-in high-definition color, all from the safety of a ground control station—and then be home from war in time to have dinner with their family. Unprecedented technology has led to an unprecedented situation.

With the advancement of technology that has enabled killing remotely from extreme maximum physical distance and extreme close empathetic distance, we must revisit the model in Figure 4 and reevaluate how we think about distance. Rather than just physical distance, we will consider total distance (physical, cultural, moral, social, mechanical, and empathetic) as it relates to the killer's ability to dehumanize the victim versus the overall impact on the killer in Figure 5.

Adapted from *On Killing*, total distance between a killer and victim consists of every factor that separates the two: physical distance, cultural distance, moral distance, social distance, mechanical distance, and empathetic distance. Physical distance between killer and victim is self-explanatory. Cultural distance includes differences in race, ethnicity, and religion. Moral distance includes the belief that one's cause is moral, just, and legal and that the enemy's is not. Social distance comes from a person's social upbringing and influences how that person thinks of other societies and classes of people. Mechanical distances are buffers that prevent the killer from viewing the victim in person, such as a screen, scope, or sight. Empathetic distance is how well the killer knows or can identify with the victim.

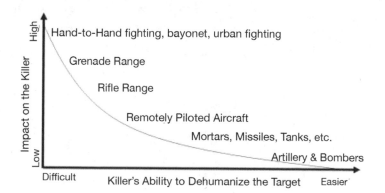

Figure 5

At one end of the spectrum, based on total distance factors, is the killer's ability to empathize and relate to the victim with respect to perceiving the victim's humanity. The opposite end of this spectrum is the killer's ability to dehumanize the victim. The larger the gap between killer and victim in each distance, the easier it is for the killer to dehumanize the target and the less the total impact may be on the killer. The total impact on the killer is a combination of how much the killer fought through the resistance to kill, their response to killing, the personal fear or danger the killer was in at the time of the kill, and the killer's knowledge of or intimacy with the victim.

How did I arrive at this conclusion of where RPAs fell out in Figure 5? To determine the killer's ability to dehumanize the target, I looked at each subcategory of total distance individually across the close through extreme-maximum-range method of killing. For each subcategory, I determined if that distance enabled the killer to deny the humanity of the victim, aided in denying the humanity of the victim, or didn't aid the killer in denying the humanity of the victim. As an example, when the kill takes place at close range, the killer and victim are in very close physical proximity (hand-to-hand combat, bayonet, knife, urban combat...). At close range, the relationship of physical proximity to the victim yields the same results as in Figure 3. The killer is close enough to understand who he is killing from a physical, cultural, social, moral, and empathetic distance. At this range, it becomes extremely hard to deny the humanity of the victim and the resistance to killing, as a result, is also very high, as is the overall impact on the killer. In this instance, the response to the kill, the intimacy with the target, and the fear and danger the killer experiences are all also on the higher end of the scale.

At the opposite end of the spectrum, at maximum range (bombers and artillery), the killer's total distance is very far; cultural distance, social distance, physical distance, moral distance, mechanical distance, and empathetic distance all enable the killer to deny the humanity of the victim. As a result of far total distance, the impact on the killer is also very low. Where this model deviates from Figure 3 is at the extreme maximum physical range with RPAs.

In the case of an RPA, the total distance falls in the middle of the chart. Physical distance is obviously very far, which enables a denial of humanity of the victim. Moral distance or the feeling of fighting for a just cause also enables a denial of humanity of the victim. Social, cultural, mechanical, and empathetic distances, however, are all factors where an RPA crew may not be able to deny the humanity of the victim.

The close social and cultural distance comes from long periods of observation during pattern of life missions and hunting HVIs. The long period of observation, coupled with the knowledge of the target and the high definition of the camera, reduces the mechanical distance to an extremely close distance. The information known about a target, such as his name, what he looks like, where he lives, what his daily routine consists of, whom he associates with, and details about his family, combined with observing the target for a long period of time with a high-definition camera, makes the empathetic distance equally as close as if the killer were in close physical proximity to the victim. The close nature of the empathetic distance may not permit the killer to deny the humanity of the victim.

How does this impact the killer when considering an RPA crew? The resistance to killing and the fear and danger experienced by the crew are relatively inconsequential. However, the response to killing and the intimacy with the target are similar to killing at close range. This creates quite an interesting dichotomy in which RPA crews may not experience a strong resistance to kill, but the impact on the killer may be significant. Holmes in *Acts of War* describes how distance doesn't provide assurance of indifference to a kill: "Even the antisepsis of distance is no guarantee that a sense of clinical detachment will prevail, and the sensations which accompany the first kill can be traumatic." Blair and House in "Avengers in Wrath" went a step further when they specifically addressed technology as it relates to RPA aircrew: "... contrary to popular myth, physical distance and technology were not mediating psychological impact—many crews were connecting more deeply with the experience, not less."

Bombing a target used to be considered impersonal. Relatively speaking, with a traditional manned bomber, it still is. Manned platforms have limited time to acquire the target, receive approval to strike, conduct a quick battle damage assessment, and return to base—all cushioned by physical distance and very little context of the target being struck. So, it should therefore follow that killing at extreme maximum range is also impersonal, not specific, and doesn't cause any distress on the part of those doing the killing. This may be true if we were referring only to physical distance. However, the more important distance to consider in this case is the empathetic distance.

Advances in technology now make it possible for us to see the humanity of the target and understand the context behind why we are targeting them. Individuals are tracked down, watched for long periods of time, and struck when the timing is right. That is as personal a kill as a sniper lying in wait for the right time to pull the trigger.

All of this personal intimacy with the victim is further exacerbated by this long period of observation. When a crew spends a significant amount of time hunting a target, they are exposed to many different facets of that victim's life. And the similarities between the victim's life and the killer's may cause great emotional responses to killing. Observing a person doing normal human activity, such as being a father and playing with his children or being a husband and embracing his wife, forms a one-sided bond between killer and victim. It's especially true if the killer can relate to the victim on a personal level, such as also being a parent or spouse themselves. Ben Shalit in *The Psychology of Conflict and Combat* reinforces that "the nearer or more similar the victim of aggression is, the more we can identify with him." The more we can identify with him, the harder it may be to kill him. If we ultimately do kill him, the greater the emotional impact it may have on the killer, regardless of the physical distance. What this may indicate is that, as technology continues to create unprecedented weapons systems and methods of killing, the most important distance to consider in future warfare may be the empathetic distance between the killer and victim.

Chapter Five

Sleep and Mental Armor

In peace and war, the lack of sleep works like termites in a house: below the surface, gnawing quietly and unseen to produce gradual weakening which can lead to sudden and unexpected collapse.

—Major General Aubrey Newman, *Follow Me*

Tripp and His Spiders

In January 1959, a radio disc jockey named Peter Tripp set out to raise money for the March of Dimes through a publicity stunt by staying awake and continuously broadcasting his radio show for two hundred hours from a makeshift glass-booth studio in New York City's Times Square. He was relentlessly monitored by doctors and scientists throughout, taking full advantage of this unique opportunity by treating the stunt as an experiment to see how the human body responded to extreme sleep deprivation. It was believed to have been the longest period of sleep deprivation recorded and documented at the time.

The results were remarkable. Tripp, normally a jovial and likable person, noticeably changed throughout the two hundred hours. By day three, his upbeat mood transformed as he became irritable and abusive, even cursing out those around him, including some longtime friends. His body temperature decreased over time and, as it lowered, he became more mentally unstable. By the fifth day, he began hallucinating. Most of these hallucinations occurred every ninety minutes in the nighttime, in what would otherwise have been a period of sleep.

That was when the doctors discovered that Tripp's brain waves showed that even though he was awake, his hallucinations coincided with his body's normal REM cycle of sleep, the period when we dream. It was as if he were dreaming while wide awake. Tripp's hallucinations included spiders coming out of his shoes, a suit of "fuzzy worms," cobwebs on people's faces, and mistaking a doctor for an undertaker coming to take him away.

By the end of the experiment, Tripp was paranoid and delusional. When he finally rested his head on a pillow, 201 hours had passed since he had last slept. Tripp's stunt indubitably changed psychologists' understanding of sleep deprivation, but it wasn't without a significant toll on his life. His wife claimed he never returned to the man he was prior to the event and it ultimately cost him his marriage.

Training to Be Tired

Sleep deprivation and its effects on the human body have been studied in great detail since the days of Tripp's stunt. However, despite all the collective knowledge learned about the perils of sleep deprivation in the last sixty years, some of our most dangerous professions haven't altered their methods of training or operations significantly.

Consider our warriors in the armed services. A majority of the skills that warriors need to survive in combat are introduced in training. The conditioning and repetition in a controlled stressful environment builds confidence in warriors that they can overcome that situation when they encounter it again. It also gives them courage to push past fear in dangerous situations, relying on their training to get through it. But even training has its limitations.

Military training rarely prepares warriors for the sleep deprivation and cumulative fatigue associated with combat with a few exceptions. One particular military training school that does a phenomenal job of pushing a warrior's body up to and beyond what one thinks is physically and mentally possible is SERE school.

SERE (pronounced "seer") is an acronym for Survive, Evade, Resist, Escape. SERE training is specifically designed to train those service members who have the potential to find themselves behind enemy lines and vulnerable to becoming a prisoner of war, most notably for aircrew or special operations forces. Each branch of service conducts the training a little differently and in varying environments, but one thing is uniform: It is some of the most challenging, intense, and realistic training our service members receive.

The premise of the training is to instruct students in a classroom environment in each portion of survival: isolation, evading the enemy, resisting torture and interrogation, and escaping captivity. Then the students are taken to the field for practical application. The practical application includes extremely harsh environments and lack of food and sleep for days. It starts with a survival and evasion stage. At some point all students are captured and become "prisoners," at which point they practice techniques they have been taught in a realistic prisoner of war scenario. This is where the real sleep deprivation sets in. I know this firsthand from my experience as a young Marine captain attending SERE training in Maine:

> Maine in the winter, as you would expect, was a fairly harsh environment to learn survival skills. There was three feet of snow on the ground and it got down to twenty degrees below zero at nighttime. I had never experienced temperatures that cold, nor had I ever slept outdoors in three feet of snow. Needless to say, the environmental conditions made for some very restless nights during the survival and evasion stages of training.
>
> By the time I was captured and thrown into the prison camp for the

resistance and escape stages, I was exhausted from evading capture in snow-shoes on a mountain in deep snow. But there was no respite for me (the senior officer among the "prisoners") or anyone else for the next fifty hours.

As prisoners, we were not permitted to sleep during this period. Loud music, constant interruptions by guards, starvation, light and time deprivation, seclusion and isolation all played tricks on our minds and kept sleep at bay. Even if we tried to sleep, a guard would wake us up in the politest way and administer some punishment for our noncompliance of the rules.

By my third day of sleep deprivation I began to hallucinate. The first hallucination was an eyeball constantly watching me through my cell door's peephole. This was most definitely a defense mechanism since the prison guards would routinely check on us through this same peephole in the door and I could see their eyeball when they did.

By the fourth day, a guard had dropped a broken piece of a mechanical pencil right outside my cell and within my reach. At this point I was starving and hallucinating frequently, and I thought that this sliver of plastic on the floor was a Jolly Rancher. With all of the stealth that I could muster, I reached under the door and snatched that "Jolly Rancher" into my mouth.

I enjoyed the sweet taste of plastic mechanical pencil for probably six hours until I realized that it was the longest-lasting and worst-tasting Jolly Rancher I had ever had and carefully put it back where I found it. All the while, hoping I hadn't fallen prey to some "Jolly Rancher trap" that would invoke extra attention by the guards.

I had never experienced this level of sleep deprivation in my life before, despite three previous combat deployments. What struck me most was how quickly my mind turned to mush from lack of sleep. I couldn't concentrate. My vision alternated between near and far, blurry and clear. I was delirious. I began to tell myself stories and laugh out loud at them. Strangely, I eagerly awaited the next playing of Rudyard Kipling's poem "Boots" over the loudspeaker in my cell so I could sing along with it. In that physical and mental state, I would have been worthless in combat or anywhere else for that matter.

Psychologically you cannot teach your body to get by without air, water, food, or *sleep*. But *psychologically*, if you must go without air, water, food, or *sleep*, and you have done so in the past, there is a period of time (before the body psychologically, mechanically fails) when you will be less likely to give way to panic or despair and can (to varying degrees) continue to function.

For example, in all military "scuba schools" divers are taught to go without air, using various stressful and generally unpleasant techniques. And the graduates of these schools really can function right up until they become unconscious, whereas the average person would have panicked and inhaled water long before that point.

In a controlled training environment, there is a place for sleep deprivation and the "inoculation" process that it can provide. But in real-world life-and-death operations, sleep deprivation represents profound leadership failure. And the primary lesson we should take away from training like SERE school is *never* to let our warriors operate when they are sleep-deprived.

You simply cannot train your body to overcome its need for sleep. Eventually it catches up to you and you start to make mistakes. It is like trying to operate your car without gas or your phone without a charge. But not all of our functions degrade at the same rate when we are sleep-deprived. Belenky et al. described this corollary in "The Effects of Sleep Deprivation on Performance During Continuous Combat Operations":

> Sleep deprivation degrades the most complex mental functions, including the ability to understand, adapt, and plan under rapidly changing circumstances. In contrast, simple psychomotor performance and physical strength and endurance are unaffected. For example, a soldier can shoot as tight a cluster of rounds at a fixed target after 90 h [hours] without sleep as he or she can when well rested, but if he or she has to shoot at targets that pop up at random at random locations on a firing range, then his or her performance drops to below 10 percent of baseline.

Sleep: The Battery Charger

How important is sleep? For your health and well-being, it is as essential as breathing. The average person needs between seven to nine hours of sleep a day. During this period, the body heals itself, recharges the battery, and backs up the hard drive. When we sleep, new pathways are formed between neurons in our brain that help us retain new information we have learned, while simultaneously the immune system fights off harmful viruses and bacteria.

If a charging phone is interrupted before it is fully charged, at some point throughout the day that phone will run out of battery life. Your body acts in a similar manner; it doesn't like to start the day with less than a fully charged battery. And your body lets you know it. The mind is an amazing regulator of the body. If you deprive the mind of the sleep it needs, it has ways of reminding you during the day that it needs a charge, such as yawning, feeling groggy, or falling prey to micro-sleeps (those little head-bobs or periods where we zone out for a second, not even realizing that we were asleep).

When the mind is working with a deficit of sleep, it works at a slower rate. This affects our ability to concentrate, think clearly, solve problems, learn and retain new information, and ultimately our body's physical coordination. If the brain is running more slowly, it processes and re-sponds to information and stimuli more slowly. This can lead to mistakes on the job and increased risk in routine tasks such as driving a car. When those tasks involve life-and-death decisions, as they do in combat, the last thing you want is a sleep deficit resulting in slower reaction and response times and degraded cognitive ability.

After eighteen hours without sleep, your body is the impaired equiv-alent of being legally drunk, a .08 blood alcohol content. This is an example of acute fatigue. We don't let our warriors engage in combat while drunk, but we seem to have no qualms about letting them fight when they are sleep-deprived. Lack of sleep may also affect a person's mood, making them irritable, moody, impatient, and more inclined to behave impulsively.

Chronic sleep deprivation can have significant long-term effects on health, including increased risk of obesity, diabetes, heart disease, heart attacks, strokes, a weakened immune system, depression, hallucinations (yeah, a Jolly Rancher!), anxiety, suicidal thoughts, and even suicide. Sleep deprivation also makes us more predisposed to become a stress casualty. When we are sleep-deprived and in a profession that involves the taking of lives, that is a dangerous combination for many reasons. Sleep serves to reinforce a person's mental armor, shielding them from stressful situations and preventing trauma from seeping in. Fatigue, acute and chronic, is a serious chink in our mental armor.

Mental Armor: The Mind's Fortress

What is mental armor? Think of it as a mental buffer that allows you to adapt to dynamic environments and increasing demands both physically and mentally. It is the mind's fortress. When your mental armor is chipped away, you are no longer able to successfully provide that top-down, executive control from your forebrain that allows you to rationalize, regulate, or adapt to your environment and the demands of that environment. When stripped of your mental armor, mental functionality is overrun by your more visceral impulses and basic instincts. Cognition declines and sensation rises.

Think about any time you've been tired, really tired, exhausted even. In that moment how well were you able to sense everything around you but maybe not make sense of it? The brain fog set in, not allowing you to think clearly; maybe you didn't remember names of objects or people or have fully formed thoughts, or you were tongue-tied when trying to speak. Now consider your mood or your emotional state at the time. Were you more irritable, inclined to escalate emotions at a moment's notice or over-react to something seemingly trivial? This is the tired state that shift work, and especially rotating shifts, eventually leads to. It is a gradual breakdown of your mental armor that leaves the mind vulnerable and unprotected.

This is a common experience for RPA warriors who work rotating shifts. It can take up to a year to fully adapt to the night shift. Every time you rotate shifts you are giving *everyone* on the crew jet lag, and all the sleep deprivation problems outlined here.

Many RPA warriors can cite instances that have manifested at home with family and friends. For example, one common complaint is poor communication with spouses such as frequently forgetting conversations, or missing important dates because they could not remember there was an event that day. Or simply just forgetting to respond to a text or phone call until days later when off-shift.

Another example frequently cited is the general irritability toward loved ones and close friends, and in some cases aggression without provocation. Most notably is the lethargy and lack of interest or willingness to do anything but sleep or keep it low-key. "I'm tired" no longer is a statement but a personality trait. Obviously, these effects are manageable as one-offs here and there, or maybe for a weekend or two at a time, but over the span of years, these warriors are no longer just "he's tired" or "she's stressed this week." These are chronically sleep-deprived and mental armor–depleted warriors who need one simple, natural reprieve: sleep.

Flight surgeons, psychological professionals, chaplains, and the like are constantly offering both their expertise and recommendations for better sleep routines, nutrition, and physical fitness. As a result, there are many RPA warriors who successfully manage shift work without it affecting them on a regular basis or causing dramatic issues at work or outside of work. Adhering to the professionals' recommendations strengthens the warrior's mental armor and enables them to not only adjust but adapt.

In addition to self-care, there are without a doubt untold benefits of support systems, of families and friends, as well as of peers and leadership, helping to buffer the imbalance. But if you're in a shitty mood and irritable because you're not just tired but exhausted, you're not exactly everyone's favorite person in the room, and likely you do not know it or cannot cognitively process it well enough, due to the mental degradation of rotating shifts and sleep deprivation.

Does that mean we have exhausted RPA aircrews killing people? Yes and no. As sleep-deprived as an RPA warrior can be, at the moment of need adrenaline kicks in. The JTAC issues a nine-line and it might as well be an EpiPen straight to the heart. This, too, is the common narrative expressed by RPA warriors. Tired or not, the adrenaline rushes in and boosts not only cognition but buffs up the mental armor. But, just like the fight-or-flight response or coming down off a caffeine or sugar rush, what follows is a low that creates a whole new level of mental vulnerability, allowing trauma to seep in: the parasympathetic backlash.

This should not be surprising. Imagine you just witnessed a traumatic event. When your mental armor is down, your higher-order thought and ability to cope is lessened. This prevents the mind from being able to properly filter and appraise all aspects of the event, and from being able to preserve a natural, balanced, and positive state of mind. Our mind has a way of processing bad things and traumatic events in order to keep us in mental homeostasis, allowing us to continue functioning and thriving. When it cannot, psychological and consequently physical problems may arise. Long shifts, and especially rotating shifts, put the RPA warrior in that vulnerable position to begin with.

The Worst Thing We Could Do for Sleep: Shift Work

Some of the most trusted professions responsible for making life-and-death decisions on a routine basis are operating on a sleep deficit: doctors, law enforcement officers, truck drivers, and our warriors in the service. These are the professions that need to be well rested and at their peak performance when executing their job. But they aren't in most cases due to chronic fatigue resulting from rotating shift work or working long shifts, or, in the case of RPA crews, both.

U.S. Air Force RPA crews routinely work twelve-hour shifts, rotating between days or nights every six weeks. Days off within the week also rotate frequently. Air Force Major Johnny Duray described how this looks for the typical RPA crew member in "Forever Deployed":

Scheduling limitations result in their [RPA crews'] "weekend" only falling on a Saturday and Sunday about once every six weeks. Once every four and a half months that weekend coincides with a normal "day-shift" circadian rhythm.

First, let's look at the dangers of long shifts. Long shifts inevitably lead to sleep deprivation. It's a simple math problem. There isn't enough time in the day to accomplish everything we want or need to, work a twelve-hour shift, and get eight hours of sleep. It is not feasible day in and day out. What typically suffers is the sleep. At the end of a twelve-hour shift a person is exhausted. They are mentally and physically drained. A Harvard study showed that twelve-hour shifts and sleep deprivation are the greatest predictors of ethical failures and use-of-force failures in law enforcement. As discussed in *On Combat*:

> Sleep deprivation is the best way to physically predispose yourself to become a stress casualty. It has been linked to mental health problems, cancer, common colds, depression, diabetes, obesity, and strokes. As it relates to job performance, sleep deprivation impairs reaction time, judgment, vision, information processing, short-term memory, performance, motivation, vigilance and patience.

A Rand Corporation study conducted in 2015 titled *Sleep in the Military: Promoting Healthy Sleep Among U.S. Servicemembers* found that a warrior's participation in combat, specifically if they experienced a traumatic event, was responsible for poorer sleep quality and frequent disturbing dreams. The study went on to state that it may not be just deployments that increase the risk of sleep problems in service members, but rather the constant conditioned vigilance and hyperarousal associated with the high operational tempo of combat. As we have seen, hyperarousal, constant conditioned vigilance, and a high operational tempo are the norms within the RPA community.

Sleep deprivation creates impaired judgment and is also one of the greatest risk factors of suicide. After twelve hours of working, people make bad decisions and terrible things happen as a result. But a twelve-hour shift is standard practice for RPA crews. And if you think crew members are smoked by the end of a twelve-hour shift, there is still a debriefing, and an hour drive home at some duty stations. This is repeated three to six times a week and over time leads to cumulative fatigue. Over the five years that RPA crews spend doing it during their first tour in the community, it can lead to exhaustion, burnout, apathy, relationship issues, and in some cases, when combined with everything else they experience, mental health issues.

The Rhythm of the Night

The second, but equally detrimental, threat to healthy sleep is rotating shifts. It takes up to a full year to adapt to a night shift, and that is if someone is not constantly rotating back and forth between days and nights every six weeks. There is nothing we could do that is more harmful to a person's sleep cycle, and subsequently their health, than a rotating shift. When a person works night shifts, he is fighting his body's natural tendencies for sleep. He is fighting his circadian rhythm. The National Sleep Foundation, the experts on sleep, described how the circadian rhythm works in "What Is Circadian Rhythm?":

A part of your hypothalamus (a portion of your brain) controls your circadian rhythm. That said, outside factors like lightness and darkness can also impact it. When it's dark at night, your eyes send a signal to the hypothalamus that it's time to feel tired. Your brain, in turn, sends a signal to your body to release melatonin, which makes your body tired. That's why your circadian rhythm tends to coincide with the cycle of daytime and nighttime (and why it's so hard for shift workers to sleep during the day and stay awake at night).

In the time before electricity was harnessed to produce artificial indoor light on a massive scale, humans' sleep/wake cycle was governed by the sun, by periods of natural light and darkness. When the sun went down, there was only fire to light the way and early hunters and gatherers would turn in for the night. Agrarian societies didn't alter this pattern much. It took Thomas Edison, and a carbonized bamboo filament found in lightbulbs after six thousand attempts of testing other materials, to alter that pattern. Once humans mastered the night, there was no regression to the days of sleeping entirely based on the sun's absence. Edison himself, esteemed inventor of the lightbulb that conquered the night, claimed to get by on three hours of sleep a night. But documents kept by his friend Henry Ford revealed Edison's secret, which Jennifer Latson described in the *Time* magazine article "How Edison Invented the Lightbulb—and Lots of Myths About Himself":

> When the Ford Motor Co. archives were opened in 1951, researchers found many pictures of Henry Ford and his pal Edison in laboratories, at meetings and on outings. In some of these photos, Ford seemed attentive and alert, but Edison could be seen asleep—on a bench, in a chair, on the grass. His secret weapon was the catnap, and he elevated it to an art. Recalled one of his associates: "His genius for sleep equaled his genius for invention. He could go to sleep anywhere, any time, on anything."

If the inventor of the lightbulb took naps during the day as part of his sleep cycle, why shouldn't we?

Sleep Hygiene: Going Caveman

Early cavemen probably slept well, tucked into a dark, cool crevice wrapped in animal furs when the sun went down. We can learn a lot from early cavemen and need to examine three things concerning good sleep hygiene: our sleep environment, naps, and stimulants and depressants.

Our sleep environment should mimic the caveman's: dark, cool, and quiet. We need the dark for our body to produce melatonin that helps us sleep. Normally this means shift workers need blackout curtains and a sleep location far from the rest of the cave dwellers. Unfortunately, we live in modern times and too often the cave gets disrupted by a common culprit: screen time. Screen time on modern devices, right up to the time we go to sleep, disturbs the cave. It negates the intent of "making a cave." Even worse is the practice of keeping the device in the room when we are sleeping because we feel compelled to check the phone whenever it lights up. Don't! Put it on *do not disturb* and get some sleep. Or, better yet, put the device in the next room when sleeping.

Where screen time is concerned, one of the worst behaviors disrupting sleep is playing video games. Video games are entertaining, but they can be addictive. It's a behavior that seems innocuous since it's enjoyable, but if it takes a negative form (e.g., overuse) the behavior can be compulsive to the point of sleep deprivation—all without you realizing it. For those who game, consider the brain fog the next morning as well as the restlessness through the night. The solution: play in moderation and only during normal waking hours.

Video games are designed to put us in a "flow state" in which it is impossible to keep track of time. Each generation of video games becomes more immersive and addictive. Suddenly it is 3 a.m. and you have no idea where the last six hours went. If you stagger into the house drunk every night at 3 a.m., you know you have a problem. If you stagger into work sleep-deprived every day because you played video games all night, you know you have a problem. If someone came to work drunk, you would respond very harshly. If someone comes to work sleep-deprived because they played games all night, they need to be dealt with harshly.

Research tells us that video games are responsible for 15 percent of all divorces in America. We have to decide now: What is important? Is our vow of marriage important? Is our oath to our nation as a service member important? Are our job, our health, and our family important? When you look at it through that lens the decision becomes pretty easy.

Go to bed and wake up at the same time each day, including weekends. Unfortunately, this is near impossible to do for rotating shift workers. One efficient sleep aid is white noise to drown out other sounds that may keep you awake: a fan, a white noise machine, but not the TV. And, lastly, treat the bed as the place where you sleep. Don't use it as the living room and dining room. These simple steps will improve your sleep cave.

Another method to combat sleep deprivation, especially for shift workers, is napping. As Edison demonstrated, naps are helpful to recharge the battery. But the nap needs to be a minimum of twenty to thirty minutes to be effective. A ten-minute nap makes us relive the startle response and does the body no good. For this reason, we should lay off that snooze alarm. It isn't restful sleep. When the alarm goes off, get out of bed. Train the body to do that and it won't need those useless nine-minute snoozes. If you have enough time to snooze three times, just set the alarm for later and get up when it goes off. Napping too close to sleep time is also disruptive to the cycle.

Caffeine is our friend, nicotine and alcohol are not where sleep is concerned. But only caffeine in moderation. Caffeine should not be the replacement for sleep. Coffee appears to be one of the best things we can put in our body. Having one or two cups of coffee for breakfast and lunch is perfect. But caffeine has a half-life of five hours, which means if you are having that last cup of coffee five hours before bedtime, half the caffeine is still in your system when you lay your head on the pillow. Consuming large quantities of caffeine makes it harder to get quality sleep, and sleep is that little vacation we take at the end of the day. Caffeine disrupts that. Why would you want to disrupt that? Unfortunately, we are in the middle of an epidemic of caffeine abuse in an attempt to compensate for our sleep deprivation. If you think you're not affected, try this. Take a day off from caffeine—if you get withdrawal symptoms, you have a caffeine problem.

Energy drinks are all bad. In 2018, two studies on energy drinks reinforced just how harmful they are. As further evidence, energy drinks are now banned from being given to our troops by the Department of Defense. The first fifteen years of war in Iraq and Afghanistan that was not the case, and we passed out energy drinks like water to keep warriors wired and awake. It was a horrible decision. In an academic environment, students consuming the most energy drinks are getting the worst grades. In a tactical environment, the ones pounding the most energy drinks are the ones most likely to nod off on the job. Consuming large quantities of energy drinks is also a major predictor of PTSD. Consider it condensed poison. An energy drink gives us only a one-hour burst of energy—subsequent drinks give us even less—and then there is a crash. Water, Gatorade, coffee, and tea are all better options than energy drinks.

I'm Just Tired

Among our warrior class, it is common practice to say, "I'm just tired" when asked how we are doing. It is a natural response because we really are tired. Stating that we are just tired is an acceptable way to complain in the military or in a fast-paced, overworked society. Our peers, family, and friends understand being tired and do not ask more. But "I'm just tired" is also a mask used when other issues exist beneath the surface. It could be a first indicator that there is a chink in the mental armor allowing traumatic experiences to seep in.

Quality sleep is immensely important to our body. There are myriad distractions that demand our attention in today's frenetic, digitally connected society. We attempt to overcome a lack of sleep in unhealthy ways that actually exacerbate the problem. Don't sacrifice sleep for those distractions. Your mental armor, health, performance, and fellow warriors' lives depend on it.

Chapter Six

Demands of Authority: Everyone's in the Cockpit

> Must the citizen ever for a moment, or in the least degree, resign his conscience to the legislator? Why has every man a conscience then? I think that we should be men first, and subjects afterwards ... The only obligation which I have a right to assume is to do at any time what I think right.
>
> —Henry David Thoreau, *Civil Disobedience*

RPAs: The Weapon of Choice on Today's Battlefield

The demand by authority figures to use RPA to kill has been on the rise since unmanned aircraft were first armed. RPAs have become the preferred weapon on the battlefield in many instances due to their low collateral damage weapons, precision, persistence, and connectivity to a support network. As a result, the RPA community has conducted a significant volume of strikes over the last two decades. The Bureau of Investigative Journalism provides one of the most comprehensive estimates of RPA strikes per year, since the government does not disclose the actual number of strikes conducted. According to the bureau, there were approximately 5,887 RPA strikes in Afghanistan alone from January 2015 to January 2019. My survey results back up these numbers.

One-third of my survey participants had participated in twenty-five or more strikes in their career, and 17 percent of them had been involved in fifty or more strikes. It is evident: the RPA community is doing a substantial amount of killing, and it has almost become an expectation when requesting an armed RPA's support. One former Predator pilot who had nine strikes in nine straight days told me of the constant demand for him to kill: "There is no job in any other field where you have a higher probability of going to work and taking a life than being an RPA pilot, not even an executioner at a prison for death row inmates." That constant demand from authority figures to kill without respite, however, may well take a toll on people.

Performing Under Pressure: Big Brother Is Always Watching

RPA aircrews' actions are openly observed; not by the civilian public, obviously, due to the secretive nature of missions, but witnessed by the "customer" or authority figure. JTACs watch RPA video feeds on a remote receiver in the field—the same video that's in an RPA's cockpit (the ground control station). Commanders can watch the same video in their operations centers, including the Combined Air Operations Center (CAOC) where an air war is controlled.

In fact, the crew never actually knows where their video feed may be playing or who may be watching. But they must always assume that there is an audience. For this very reason, one of the largest operational stressors that a crew experiences is performing under pressure for an audience. In nearly every interview I conducted, crew members divulged this repeatedly, regardless of service or crew position. RPA crews are afraid of screwing up while an audience watches. It is not widely spoken of, but the impact of the constant pressure to perform under relentless observation should not be underestimated. Especially when that audience includes several authority figures. Many manned aircraft carry similar sensors that broadcast video feeds, but the audience is limited. There is no military equivalent of an RPA video feed, and it stresses out our crews.

Traditionally, aviators have flown in a sterile cockpit free of external distractions, with only the essential information needed to aviate, navigate, and communicate allowed into the cockpit. With the separation of cockpit and aircraft, as is the case with RPAs, this tradition of maintaining a cockpit free of distractions has eroded. Despite this downward trend, the requirement to aviate, navigate, and communicate in a distraction-free environment remains, regardless of the location of the cockpit. Unfortunately, this one design shift has rendered RPA cockpits more accessible in a number of different ways. Due to the networked and connected nature of the ground control station, authority figures can chat with or call directly to RPA crew members while simultaneously watching their video feed. A prime example: During Scott Swanson's first kill with a Predator in 2001, the commander of forces in the Middle East, General Franks, watched live and provided real-time orders of what to do. In a sense, from day one we have not honored the tradition of letting the crew fly their mission uninterrupted. Instead, everyone is in the cockpit now, including authority figures, and that seemingly insignificant detail actually has a consequential impact on the RPA community.

Milgram and the Influence of Authority Figures

What do we do when authority figures are watching? Does it influence our behavior or change how we would normally perform our job? Absolutely. The Yale University psychologist Stanley Milgram performed the most famous experiment on the subject of obedience to authority. Milgram's research question resulted from the attempt of many German soldiers to use the defense of "I was just following orders" to justify their actions in the Holocaust. Milgram wanted to know if the desire to obey the orders of authority figures overshadowed an individual's moral compass. Could large masses of soldiers be ordered to commit atrocities, and would they comply even if they knew what they were doing was morally wrong?

For Milgram's experiment, he selected volunteers to serve in "teacher" and "learner" roles. The participants, forty males between the ages of twenty to fifty with various socioeconomic status, drew straws to determine if they would serve as the teacher or the learner. However, Milgram fixed the experiment so that all the volunteers ended up in the teacher role. Milgram then hooked up the learners with a set of fake electrodes that gave the teachers the impression they could shock the learners. All of the learners, who were plants working with Milgram, were then separated from the teachers in an adjacent room. In the same room with the teachers was an "experimenter" wearing an official-looking gray lab coat and holding a clipboard, in this instance representing an authority figure.

The learner was to study a set of word pairs and memorize them. Then the teacher would quiz the learner on a word and evaluate the response. Each time the learner answered the word pair incorrectly, the teacher was to administer an electrical shock through a switch on the table in front of him, ranging from a slight shock of fifteen volts up to a fatal shock of 450 volts.

Milgram designed the experiment so that the learners gave mostly wrong answers and the teacher would have to administer a shock. Milgram wanted to see just how far the teachers would go. Of course, the learner was not actually being shocked, but really sold it well from the other room by making sounds of distress and pain as if they were getting shocked. If the teacher refused to continue administering shocks, the authority figure would intervene with a series of increasingly demanding prodding statements:

Prod 1: Please continue.
Prod 2: The experiment requires you to continue.
Prod 3: It is absolutely essential that you continue.
Prod 4: You have no other choice but to continue.
Etc.

As a result, 65 percent of the teachers administered fatal shocks to the learners and 100 percent of the teachers administered a harmful shock up to three hundred volts. Milgram concluded that people would carry out orders from an authority figure, even to the point of killing an innocent human being, because of human conditioning from childhood to follow orders. He described this in the article "The Perils of Obedience":

> Stark authority was pitted against the subjects' [participants'] strongest moral imperatives against hurting others, and, with the subjects' ears ringing with the screams of the victims, authority won more often than not. The extreme willingness of adults to go to almost any lengths on the command of an authority constitutes the chief finding of the study and the fact most urgently demanding explanation.

Milgram concluded that people behave in one of two ways in social situations: an autonomous state, where the individual controls his own actions and accepts responsibility for his actions, or a subordinate state (which Milgram termed an agentic state), where the individual carries out actions on behalf of an authority figure and passes the responsibility and consequences of the actions off to that authority figure. Milgram continued that in order for someone to be in an agentic state, he must recognize the authority figure as legitimate and must believe that the authority figure will accept responsibility for the outcome of the actions.

Milgram's agency theory obviously played out during the experiment. Many teachers in the experiment only continued shocking the learners when the authority figure intervened to state that he would accept responsibility for the outcome. At that point, the teachers switched from acting in an autonomous state to an agentic state and continued shocking the learner, with nearly two-thirds of them administering a lethal shock. Now, how does this apply to trained and conditioned warriors, particularly RPA warriors?

Killing on Demand: The Impact of Authority Figures

Throughout the history of warfare, we have observed how the presence of an authority figure on the battlefield steels the nerves of the warriors and enables them to overcome their resistance to killing. At least, that is the traditional perspective on authority figures or leaders in the military: a person of courage and strength of character who rallies the troops and changes the outcome of the battle. While that occasionally happens, what predominantly happens is that soldiers do not want to appear to be disobedient in front of authority figures or incompetent and cowardly in front of their peers, and they are motivated by a fear of failure more than a fear of the enemy.

As discussed in *On Killing,* only 15 to 20 percent of frontline troops in World War II fired their weapon at the enemy when given the opportunity. When a leader was present issuing orders to fire, nearly everyone complied. Sigmund Freud understood this when he said, "Never underestimate the power of the need to obey." Crew-served weapons, by contrast, were usually fired because of the anonymity of fighting as a crew and the strength of violence in numbers. Conversely, when a leader was not present, the preponderance of soldiers, on all sides, did not kill. So how does this translate to RPAs? Consider them a crew-served weapon with an omnipresent leader.

Our modern-day warriors are the best-trained soldiers that history has ever seen. RPA crews receive training on rules of engagement, the law of armed conflict, and the mitigation of collateral damage when striking a target. Additionally, they receive volumes of training on capabilities and limitations of the aircraft, and tactics, techniques, and procedures for employing their weapon system.

Modern-day warriors are also ethically aware. Some services have guided discussions on the ethics of killing in entry-level schools or in units. In the Air Force RPA pipeline, discussions on killing occur early in the training as a method of weeding out those officers who initially know they will have a problem with killing and as a reality check for those who have never thought about killing with an RPA. This screening

231

and training ensure that the best, most capable RPA crews are at the controls.

Despite all this training, RPA crews do not normally engage targets independently. They do not make the decision to kill. In fact, they nearly always strike targets at the behest of someone else, or, according to the Milgram model, they are always acting in an agentic state. There is an authority figure who requests them to kill on his behalf. The Air Force RPA pilot Dave Blair and the psychologist Karen House in "Avengers in Wrath" described how this might affect crews: "The decision to authorize a strike is made well above the level of the aircrew, but if they cannot reconcile themselves to the logic of that decision, it becomes much more difficult to bear." This is almost a perfect definition of *cognitive dissonance*, which we have discussed as a factor contributing to a negative or traumatic response to killing.

The Context of the Demand to Kill Matters

Most times the authority figure requesting a strike is a Joint Terminal Attack Controller, who is acting on behalf of a commander on the ground in the combat zone and who is in direct communication with the RPA crew. When the JTAC declares that a TIC (troops in contact) situation exists and requests that an RPA strike, oftentimes the JTAC is taking fire from the enemy. The prods from the authority figure in these instances are intense and usually more colorful than in Milgram's experiment, which can add to the seriousness and timeliness of the request to kill. Put another way, when the authority figure is in danger, he emphatically demands that the RPA crew make the threat go away quickly and violently.

Despite the intensity of this request, most people respond well to it. As my research revealed, directly supporting friendly forces on the ground only caused psychological problems for the crew if, during the course of that support, friendly forces were wounded or killed. Killing enemy combatants attacking friendly forces actually produces positive responses:

relief, a sense of mission accomplishment, and pride in success at saving friendly forces' lives.

This should come as no surprise. Striking a target in defense of friendly troops is an easy situation to rationalize and justify. In this instance, the demands of the authority figure are inconsequential when the mission goes well. Conversely, the demands of the authority figure to kill on his behalf may have a negative compounding effect on the crew when the mission results in friendly casualties. "Why couldn't I save them?" "If only I would have been a little faster, a little more proficient." This situation may create survivor guilt because the crew does not share the same risk as the troops they support remotely.

When the authority figure demands that an RPA strike a target, knowing there will be civilian casualties as an outcome, the situation becomes more complex to resolve in the warrior's mind. In a perfect war, there would be no civilian casualties. Unfortunately, we do not fight perfect wars. We live in a world where HVIs use civilians as a shield to obfuscate the rules of engagement, and that greatly complicates matters.

Despite this tactic, HVIs surrounding themselves with noncombatants does not always work out in their favor or curtail the demand of the authority figure. For example, depending on the importance or tier of the HVI, authority figures may deem it necessary to accept civilian casualties when the fleeting opportunity to kill a top-level HVI occurs. Irrespective of the moral dilemma of killing one extremely bad actor and innocent civilians, there is as a trade-off: the prevention of all the sinister activity that bad actor may perpetrate in the future if we do not kill him when the opportunity presents itself. This is very rare. It is not common practice. The U.S. military goes to great lengths to eliminate civilian casualties wherever possible. These RPA crews are rooted in their knowledge of history, and there may be some peace to be found in comparing their actions to World War II bomber crews who wiped out entire cities of civilians in their endeavors to defeat the enemy. Or artillery crews across the centuries who knew that their cannon fire was very often killing civilians. But in these historical cases, the bomber and artillery crew did not have to witness the harm they were inflicting.

The majority of the time, RPA crews let the HVI get away rather than kill nearby civilians. Again, there are times when it becomes necessary to kill an HVI immediately, regardless of the surrounding circumstances. When that happens, the authority figure making the decision to kill is elevated to higher levels based on the importance of the HVI and the number of estimated civilians who may die because of a strike. Essentially, each level weighs the predicted gain versus loss while still exploring alternatives. This decision process extends to some very high-level leaders, up to and including the commander in chief (the president). Talk about demands from authority.

In that situation, the RPA crew is acting in an agentic state on behalf of the supreme authority for the military. Where does that mentally leave the RPA crew? Potentially killing civilians because someone both authorized them to and accepted responsibility for it on their behalf. Legally, the crew would have engaged the target meeting the rules of engagement. Ethically, they may feel like they just amped the voltage up to 450 and hit the switch because the person in the gray lab coat said "you have no other choice but to continue" and that he would accept responsibility for the outcome. Fortunately, our ethical warriors cannot just shed their morals; situations like this may lead to cognitive dissonance, where what the crew is required to do differs from how they morally feel about doing it.

The context of the demand from an authority figure to kill on his behalf is critical. When crew members do not understand why they are killing a target, it may start to sow a seed of doubt in their mind about the entire system. That doubt erodes a crew's trust in the authority figure who requested them to kill, erodes trust in the authority figures within their chain of command who put them in that position, and ultimately erodes the trust in the legitimacy of authority figures. The doubt may make the crews question why they are killing certain individuals. The killing may not seem justified in their mind unless the context of "the why" behind it renders the act conscionable.

Killing without an understanding of why may have a significant negative impact on the mental health of crews. Air Force Major Alexander "Epic" MacPhail provided one example of how a context void affected the people in his squadron. MacPhail at the time of the interview was the director of operations (DO) for an MQ-9 Reaper squadron. During his tenure as the DO, the missions his unit supported shifted in nature.

Initially his squadron supported the fight against the Islamic State in Iraq and Syria (ISIS), a fight in which MacPhail said his squadron was proud to participate. They were striking a significant amount of ISIS fighters during that time. ISIS was a brutal enemy who had no compunction about murdering women and children and beheading those they captured; it was, by any account, an easy enemy to justify killing. In other words, the context of why ISIS was being killed was well known in the squadron and there were very few, if any, people who had an issue with it.

However, MacPhail's squadron shifted to supporting operations in another theater, where they were striking mostly HVIs instead of ISIS foot soldiers. The squadron rarely observed these new targets conducting any hostile actions or nefarious activities. As a result, MacPhail said that the mood within the squadron changed for the worse. More people within the squadron started questioning why they were killing these individuals and more people started having difficulty with what authority figures were asking them to do.

This shift in mission resulted in more crew members seeking care from mental health professionals. In both squadron missions, the targets were valid, legitimate military targets, but the context was missing behind the second mission. As Blair and House stated in "Avengers in Wrath," in this situation, in the crew's mind, they go from a "just kill" to "just killing." It is not hard to see how if you are just killing with no context, this could affect you mentally.

Why would there be no context? Why wouldn't we arm the crew with as much information as possible about the nature of the targets? One MQ-9 pilot who had been flying operational missions for only seven months shared his experience:

Being in the pilot's seat you are only privy to some of the information about the strike, partially because of time and because there is a huge infrastructure behind it. Sometimes it seems from my perspective that it is arbitrary. Sometimes it seems like we will just find random people with nefarious activity or some items on them and we'll decide that we need to strike them. I've seen a few of those and I questioned what the reason of the strike was.

Rest assured that this is not the case, but this is occasionally how the aircrew—particularly the less experienced crew members among them— can feel about the situations they find themselves in. Due to the fast-paced nature of combat, there is not always time for a crew to learn more about the nature of the target. And this level of context is not an absolute requirement for them to be able to justify the killing. As an example, defending friendly troops in contact with the enemy is easy to justify without knowing anything about the target other than that he is trying to harm friendly forces. That is easy to rationalize and makes it a just kill.

In contrast, when the target is a high-value individual and there is no context behind the strike, the crew may not be able to rationalize the demand from the authority figure to kill on their behalf and they may feel like they are just killing. In some cases, there is no time to get information to the crew. Sometimes it is a classification issue: an authority figure makes the decision on the crew's behalf that they do not have a need to know why they are striking a target.

These situations require an extreme amount of trust to overcome. When trust between intelligence and operations has not been established, it may lead to circumstances such as described above by the inexperienced MQ-9 pilot, who had a biased and incomplete understanding of how the "system" reaches the decision to kill an individual and selects them to do it. The need for a relationship based on truth cannot be overstated here. Context and trust are vital for RPA crews while conducting this most morally challenging aspect of the job. The difference between a "just kill" and "just killing" in the crew's mind is germane to our discussion on the human responses to remote killing and could be the difference between trauma and absolution.

The Decision Not to Kill: Moral Victory

Very few people I interviewed would talk about missions they were involved in that resulted in collateral damage. Just as discussed in *On Killing*, though, the warriors were very eager to talk about the times when they did not agree with a situation and chose not to kill. From *On Killing*:

> On rare occasions those who are commanded to execute human beings have the remarkable moral fiber necessary to stare directly into the face of the obedience-demanding authority and refuse to kill. These situations represent such a degree of moral courage that they sometimes become legendary. Precise narratives of a soldier's personal kills are usually very hard to extract in an interview, but in the case of individuals who refused to participate in acts that they considered to be wrong, the soldiers are usually extremely proud of their actions and are pleased to tell their story.

One of those legendary moments occurred because of an RPA crew refusing to kill an American citizen. Anwar al-Awlaki was a U.S. citizen born in New Mexico and was a Tier 1 target based on his affiliation with Al Qaeda. He was a Yemeni-American preacher and the leader of Al Qaeda in the Arabian Peninsula in Yemen. The American government wanted al-Awlaki captured or killed for a laundry list of justifiable reasons, despite his American citizenship. Al-Awlaki colluded with the Army psychiatrist Nidal Malik Hasan, who killed thirteen people in a mass shooting at Fort Hood, Texas, was involved in the planning of the failed Christmas Day "underwear bomber" on an airplane in 2009, declared jihad against the United States, and constantly plotted to kill Americans and foreigners.

Al-Awlaki may have been born in the United States, but his loyalty was to a radical extremist group bent on killing Americans. He was a domestic terrorist living abroad, which presented a legal and moral quandary. Normally this kind of situation would be a law enforcement issue, but al-Awlaki was deeply rooted in the hinterlands of Yemen, a relatively lawless part of the country where there was little possibility of arresting or capturing him at the time. Based on the challenging circumstances, the American government decided to target al-Awlaki with an RPA. Due to al-Awlaki's citizenship status, this decision was controversial and not everyone agreed with it. An RPA pilot within a special operations squadron recalled to us the story of his unit refusing to strike an American citizen:

> I was on at Christmas and there was a red phone next to us and I was told that if it rings it will be Obama who will authorize you to take the shot. We had been following this guy for months and months. It was in a theater that was not known at the time that we were even operating in [later known to be Yemen]. Since this guy was an American citizen, people had a problem with it within the squadron. We thought it was illegal and we refused to do it. When the leadership told them to shut up and color [follow orders], that commander was dismissed and that line was taken away from us and given to the CIA.
>
> They had different rules to fight under. People were concerned the military was targeting a U.S. citizen. I thought it was great moral restraint exercised by our officers of that squadron in making sure that target set was removed from the deck no matter how badly national leadership wanted this target taken out. To do that to an American citizen was a moral impasse.

Anwar al-Awlaki was ultimately killed in an RPA strike on September 30, 2011, in Yemen. He was the first American citizen known to be deliberately targeted and killed by an RPA. It is a policy question whether al-Awlaki should have been killed and pertinent to our discussion only because of the uncharted, complex situations our RPA crews find themselves in when they are killing. The president of the United States, the commander in chief, directing an RPA crew to kill a terrorist who happens to be an American citizen is quite possibly the greatest moral dilemma one could face as an RPA crew. We are well beyond experimenters in gray lab coats as authority figures.

Missing the Shot: A Warrior's Protest to Killing

We know that even when an authority figure orders warriors to fire, the warriors can simply exercise their right to miss or fire over the enemy's head to counter the demands of authority and simultaneously avoid their reluctance to kill. In most of those instances, it is unknown by the authority figure whether a warrior missed and if they did it deliberately.

This still happens with RPAs. What is different with RPAs, however, is that everyone watching the video knows the results of the shot. There is no hiding, no deliberate misses to avoid killing, and no doubt about the outcome of your shot. Gwynne Dyer in *War* stated, "Men will kill under compulsion—men will do almost anything if they know it is expected of them and they are under strong social pressure to comply—but the vast majority of men are not born killers." When the authority figure is watching your work in real time over a video feed, there is no greater compulsion to comply and our RPA warriors are no exception.

Distance Between the Authority Figure and the Killer

Milgram conducted his famous experiment with many variations. One variation, referred to as the "Proximity Series," altered the distances between the learner and the teacher. The results? Initially, when the learner was next door in a different room, 65 percent of the teachers initiated a lethal shock, but when the teacher was face-to-face with the learner that number dropped down to 30 percent. What we see in the case of RPAs is a physical separation of the authority figure, the killer, and the victim. Milgram's experiment is playing out every day in the RPA community with one major difference: RPA crews observe the aftermath of their actions on a video screen as opposed to hearing the screams of their victims next door.

Similar to the impact of the presence of an authority figure, the physical distance between an authority figure and the killer has an effect as well. They communicate with each other, but the two are never physically in the same space. In the past, the presence of the authority figure made it easier for the soldier to comply with the authority figure's demand to kill. It allowed the soldiers to transfer responsibility to the authority figure for their actions. The physical presence of the authority figure on the battle-field facilitated compliance with the order to kill. RPAs have once again upset that traditional model.

There is another dynamic at play here, though. The physical separation between the authority figure and the warrior may make it easier for the authority figure to issue the order to kill, and the physical distance between the warrior and target may make it easier to comply with the order. Simultaneously, the empathetic distance between the warrior and the target may make it more difficult to deal with killing.

The RPA model, with the physical separation of authority figure, killer, and victim, is unlikely to change. The context provided to the killer, based on the recognition of this challenging dynamic, may help with the killer's rationalization of the kill and overall mental health. Understanding this challenge, we should endeavor to provide as much context about the target to RPA crews as possible when they are killing in an agentic state on behalf of authority figures.

Chapter Seven

Group Absolution: Killing as a Crew

All crowding has an intensifying effect. If aggression exists, it will become more so as a result of crowding... The effect of the crowd seems to be much like a mirror, reflecting each individual's behavior in those around him and thus intensifying the existing pattern of behavior... thus increasing the dehumanization that "transfers men into beasts."

—Ben Shalit, *The Psychology of Conflict and Combat*

On June 18, 2010, Ronnie Lee Gardner was executed by a Utah State firing squad for a murder conviction twenty-five years earlier. In 1984, Gardner viciously killed a man by shooting him in the face during a bar robbery and later killed an attorney during a failed attempt to escape custody. Gardner led a troubled life from an early age, abusing drugs and spending time in and out of juvenile detention and prison. Following his second murder conviction, he was sentenced to death in Utah. He spent the next twenty-five years appealing the sentence all the way to the U.S. Supreme Court, which ultimately refused to hear the case.

In the end, Gardner chose to be executed by a firing squad for religious reasons, believing as a Mormon that a blood atonement for his sins was needed for him to receive forgiveness from God for murdering two people. This was a very high-profile case widely covered by the media and, in a strange twist of events, the day prior to the execution, the Church of Jesus Christ of Latter-day Saints (commonly known as the Mormon church) released a public statement rejecting blood atonements as counter to their religious beliefs and doctrine despite some historical precedence.

But the plans for the firing squad continued and, the next day, Gardner was executed. He was bound to a chair with a hood draped over his head and a paper target pinned over his heart. Five anonymous police officers, all volunteers, armed with .30 caliber rifles, simultaneously fired a single shot each from twenty-five feet away. Four of the five officers' rifles were loaded with a live round and one was randomly loaded with a wax bullet to mimic the recoil of a real bullet, thereby creating uncertainty within the firing squad over who actually fired the fatal shots.

To initiate the execution, a backward countdown from five was verbalized, with all rounds being fired simultaneously when the countdown reached two. The four rounds successfully found their target, and Gardner was declared dead two minutes later. It was the first time in fourteen years that a prisoner had been executed by firing squad in the United States.

Killing as a Crew: Why It's a Firing "Squad"

Why is this execution relevant to our discussion on killing with RPA? Because RPA warriors kill as a crew, and much can be understood about the psychology of killing as a crew by looking at a firing squad where the killing environment is planned and controlled.

First, and probably most important, there is diffusion of responsibility and mutual surveillance in a group. Among the shooters in a firing squad you can never know which bullet actually killed the victim. Likewise, in almost every instance of a military weapon employed by a crew (machine gun, tank, artillery, aircraft, or RPA), killing is done as a group. These groups, and the weapons they employ in combat, will almost always fire.

After diffusion of responsibility and mutual accountability, the next factor enabling RPA kills is anonymity. We know that people are capable of doing things anonymously that they would never do otherwise. The identities of the members of the firing squad remained unknown and unreleased to the public, and the reassurance of anonymity serves to bolster the resolve. Similarly, RPA crews know that their identity will remain anonymous for doing their nation's bidding.

The prisoner's hood also helped to prevent the executioners from confronting the victim's humanity, particularly from looking at the prisoner's face and into his eyes prior to the kill. A hooded victim enables the killers to deny the humanity of the victim and the execution becomes more akin to target practice. Military crews employing a weapon at a distance also never have to look their victims in the eye. Using a hood ensures that the condemned never sees it coming, just as when ambushing an enemy with a standoff distance weapon such as a Hellfire missile, artillery shells, or a bomb.

Among the members of the firing squad, there is a diffusion of responsibility. Due to the introduction of the wax bullet in one shooter's rifle, there is uncertainty about each individual's level of participation and responsibility for the act of killing. Each shooter diffuses a portion of the responsibility of the kill to the other four shooters despite the fact that each had an 80 percent chance of having an actual bullet in their rifle. There is the anonymity of not knowing who actually killed. An RPA crew experiences a similar diffusion of responsibility.

The firing squad's countdown backward from five is akin to an authority figure directing them to fire. This is similar to the RPA receiving a "cleared hot" command from a JTAC, giving a verbal authorization to strike a target. In fact, JTACs "own" the bombs or missiles that they clear for a strike. In theory, this relieves striking RPA of the responsibility of the effects of the strike.

The verbal countdown forms a bond of trust between the shooters in a firing squad. It's highly likely that if one shooter "opted out" at the moment of truth the others would know. This pact to fire at the same time ensures that each shooter fires, and the fear and pressure of letting down the group helps overcome each shooter's resistance to killing. The collaborative nature and openness of how an RPA crew conducts a strike likewise makes it unlikely that anyone will "opt out" when the clearance to strike has been issued. There most certainly have been instances of dissension and refusal to kill among RPA crews when there were moral disagreements with the requesting authority. However, the majority of the time, once legal authority to kill has been attained within the rules of engagement, the crew is striking that target.

Shared hardships help formulate a bond among any group. This bond can be forged from mutually experienced extreme situations, such as combat or the execution of a condemned prisoner. In the case of the execution of Ronnie Lee Gardner, there was even a special execution coin commissioned and given to the shooters for their participation. Bonds formed during bloodshed have no equivalent comparison, and members of the group will do almost anything not to let the others down. The military equivalent, albeit not given for every strike, is a commander's coin, squadron coin, or medal. These awards help reinforce the process of absolving the act from the individual.

Killing for a legal, moral, and just cause, such as the execution of a murderer by firing squad or striking a violent extremist belonging to an international terrorist organization such as ISIS, empowers killers. In the instance of an RPA crew or a firing squad, the killers are serving the will of a higher authority, the state that has sentenced a prisoner to death, or the U.S. government that has engaged in war against a declared enemy.

But why does it have to be a firing *squad*? One well-trained marksman could easily kill an immobile, bound, and hooded person with a rifle from twenty-five feet away. Aren't five shooters a bit overkill? The key is group absolution: anonymity, shared hardship, accountability to peers, and diffusion of responsibility that enables a group to kill whereas an individual standing out there by himself may struggle with the deed. Strength in numbers is not only a concept but a practical reality in the act of killing.

The Firing Squad Analogy: Now with RPA Context

While comparing an RPA crew to an execution squad may illustrate how both groups overcome the humanity of the victim at the point when they pull the trigger, what the execution-squad comparison lacks is context to accurately describe how an RPA crew kills as a group. What follows is a more complete description of the killing landscape of an RPA crew using the firing squad analogy.

Imagine that every day the firing squad reports to work at the prison, where they sit in a windowless "watching room" that includes a large high-definition television and a phone. Displayed on this television is a hidden-camera shot of a person whom they intently watch going about their everyday life for days, weeks, months, or possibly even years. The squad is given a dossier on this person, and they soon learn that his name is Ronnie and that Ronnie is married and has three children ages eight, six, and two.

The hidden camera captures some pretty intimate moments and some very humanizing events in Ronnie's life, such as Ronnie playing catch with his children after dinner, or Ronnie embracing his wife when he leaves for work, or Ronnie shopping at the market and, yes, committing crimes that sometimes result in the deaths of other people. Routinely, while the squad watches Ronnie on TV, their boss calls on the phone and relays information on all the bad things Ronnie has been up to and the crimes he has committed during the squad's time off. The squad knows that when the time is right, they will have to kill Ronnie for the things he has done and the choices he has made.

Then one day, with little warning, they watch as Ronnie gets arrested and transported to a field where he is tied to a chair and a hood is placed over his head. The phone immediately rings and it's the squad's boss frantically telling them to get their guns loaded and come outside. It's time to kill Ronnie. When they come outside, the execution happens as described before: Ronnie is there, bound and hooded, with a target pinned to his chest, the squad stands twenty-five feet away, there's a countdown from five, a simultaneous shot, one wax bullet, etc.

Immediately after, the firing squad is rushed back into their room to watch the monitor where the camera is now zoomed in on Ronnie as he bleeds out and dies. Ronnie's hood is removed and his family is let in to the execution area to see his lifeless body. The squad now bears witness to the uncontrollable wailing of Ronnie's widow and the shock, confusion, and uncertainty on the faces of his children. And the camera keeps rolling. The fact that the squad knows Ronnie is a bad guy and has been condemned to death for his actions doesn't make killing him or rationalizing their actions easier.

The phone rings again and the boss now requests for the squad to keep watching Ronnie's body all the way through his funeral to see whether any other criminals show up who may be of interest. So the squad sets in for the long, morbid day. At the end of their twelve-hour shift, the firing squad goes home to their families. When asked how their day went, they dare not speak a word of what they have done. First off, it's against company policy to talk about details of the job and, even if it wasn't, the squad members don't want to burden their loved ones with this surreal and violent experience.

The next morning the firing squad reports back to the "watching room," where a new subject named Tom is displayed on the television. The process begins all over again and the squad knows that one day, when the time is right, they will kill Tom, too.

In our fictional scenario, the details of the firing squad's actions during the execution remain unchanged from the actual Gardner example. What changes is that we now have an understanding of the context of what the squad endures before and after the kill. The strike may be a group effort, but that doesn't necessarily make it easier to carry out. In fact, there has never been a crew-served weapon like this in history and, as mentioned, context plays an important role in killing with RPA.

Group Dynamics: Polarization and Mob Mentality

Group dynamics and group behavior are a great source of insight into human behavior, particularly when the group engages in violence and killing. Konrad Lorenz, the ethologist and Nobel Prize winner, once stated, "Man is not a killer, but the group is." We like to imagine that an organized group of individuals provides for checks and balances of one another's behavior; that ultimately the morally right and just decision will emerge from a group as a result of the expression of many diverse opinions and peer pressure to do the right thing. Group dynamics do work in this manner to some extent by helping people within a group to conform to standards and norms. This effect is called polarization, and it was described well by Emanuele Castano, Bernhard Leidner, and Patrycja Slawuta in "Social Identification Processes, Group Dynamics and the Behaviour of Combatants":

> A similar process may be at play in the context of combat. Unit members are continuously placed in highly stressful situations in which dehumanizing rhetoric about the out-group finds easy confirmation in everyday occurrences. The cohesiveness of the unit...is likely to increase and lead to group dynamics through which the group ends up with behavioural decisions that are more polarized than those of each individual member.
>
> It should be noted that polarization found in social psychological research does not suggest that group behaviour will always be more negative than individual behaviour. In fact, the term polarization is used to convey the idea that the group will be likely to reach behavioural decisions (or to form attitudes) that are simply more extreme than those of individual members. If initial attitudes and behavioural intentions are positive, the group will probably be polarized towards more positive behaviour.

Group members tend to be polarized to whichever direction the group leans, but to a more extreme level than as an individual. Group behavior is one of the reasons militaries fight as units, ranging from a two- to three-person crew up to divisions and armies. Instilled discipline and training, combined with aggression from a group, contributes to making a crew, group, or unit an effective killing force.

But just as that peer pressure to conform within the group keeps group members in check, it can also serve as a contagion for violent behavior. One negative by-product of a group's actions is what's referred to as "mob mentality," a violent behavior that rapidly spreads within a group. Simply put, people become more aggressive when in a group. And when others in the group are behaving aggressively, it becomes contagious among the group's members, to the point where individuals acting in a group will do things they would never do as an individual, including commit horrible atrocities. There are many classic examples of this phenomenon throughout history: the Salem Witch Trials, the French Revolution, the Holocaust, the My Lai massacre, and the Arab Spring. As a result of group absolution, groups are just more aggressive and more inclined to kill than individuals.

Accountability to Peers: Killing Under Surveillance

Four main factors enable the existence of group absolution: accountability to peers, diffusion of responsibility, shared hardship, and anonymity. Ardant du Picq referred to the concept of the accountability to peers as "mutual surveillance" and stated that it was the predominant psychological factor on the battlefield. S. L. A. Marshall in *Men Against Fire* described this accountability to peers as "one of the simplest truths of war that the thing which enables an infantry soldier to keep going with his weapons is the near presence or presumed presence of a comrade."

Whereas individual soldiers might exercise their "right to miss" when firing on the enemy, personnel manning a crew-served weapon are accountable to each other and the crew knows if someone missed deliberately. For this reason, crew-served weapons ranging from machine guns to artillery to aircraft are extremely lethal on the battlefield.

But it isn't foolproof. RPA warriors still find ways to exercise their "right to miss" when they haven't overcome their inhibition to kill, or when they morally don't agree with the request to strike. A U.S. Army Gray Eagle pilot told me why he thinks this happens within the Army RPA community:

> Command authority doesn't exist [in the unique circumstances of Army RPAs]. In the Air Force, the pilot is an officer and the sensor operator is an enlisted position. In the Army, both crew positions for the RPA are enlisted. They could be the same rank or the person in the seat next to you could outrank you. The aircraft commander position is a real paper tiger that dissolves as soon as we leave the box and everyone knows it.
>
> I have seen crew members refuse to fire and crew members miss on purpose, and those things I read from *On Killing* about a crew-served weapon don't typically happen with a crew-served weapon. When you are about to kill someone and they don't agree with it I have absolutely seen them refuse and miss on purpose. One hundred percent. I don't know of another service that has had that happen. It's easier for us to say that we missed because we suck, rather than the guy next to me missed because he disagreed with the shot.

Deliberately missing a shot is one form of protesting the request to kill. Another form of protest is simply to state to the requester that your aircraft has reached a fuel status that requires you to return to base (known as "bingo") and that you are unable to assist them due to safety of flight.

Although I have tended to refer to RPA crews as monolithic, it is important to note that they are not. Each service and country trains and employs RPAs differently, and there are unique challenges and differences in each distinct employment. As the Gray Eagle pilot mentioned above, if two enlisted soldiers of the same rank are on a crew, each may feel inclined to do what he wants rather than what is asked of him.

By contrast, my research and personal experience suggests this rarely applies to the Air Force or Marine Corps model, where the crew is a more traditional mixture of officer and enlisted ranks.

Diffusion of Responsibility: Who's the Killer?

Diffusion of responsibility is another of the four factors creating group absolution. Diffusion of responsibility occurs when each member in a group rationalizes that they are responsible only for a small portion of a deed, but not the overall deed itself. Crew members tell themselves that what they are doing isn't actually pulling the trigger and that someone else within the crew is more accountable for killing. This enables them to be an accomplice to the event but not bear the full weight of the act of killing.

There is a diffusion of responsibility within an RPA crew. The JTAC, MIC, or intelligence personnel make the determination of positive identification of the target (the authority to kill legally), the pilot pulls the trigger, and the sensor operator guides the missile onto the target. No single individual is responsible for everything. In fact, one pilot I interviewed dismissed the idea that he was responsible for killing since he only pulled the trigger and didn't guide the missile to the target. A British Reaper MIC stated that he was 33.3 percent responsible for everything that happened during the mission. The psychologist Albert Bandura described this diffusion of responsibility within an RPA crew in "Disengaging Morality from Robotic War":

> At the agency locus, people evade personal accountability for harmful conduct by displacing responsibility to others and by dispersing it widely so that no one bears responsibility. This absolves them of blame for the harm they cause.

Shared Hardships: Unbreakable Bonds

Shared hardships endured by a group also enable group absolution through trust. And there is no greater bond formed than with those whom you fight a war. When the group is in danger, that bond strengthens. When the environmental conditions are unbearable, that bond strengthens.

When the group is bored beyond belief, the bond strengthens. Anytime there is shared misery, the group becomes tighter.

At the most basic level, soldiers don't fight for some grandiose cause, they fight for the person to their left or right. They fight to keep the group alive, to perpetuate the bond. At prima facie it doesn't make sense that this bond would exist among RPA crews. They are not physically in danger, and the environmental conditions are controlled inside the box. They don't sleep in the dirt, or get rained on, or shot at. The person to their left or right experiences the same controlled environment and conditions.

But the bond does exist. RPA crews do endure shared hardships. Those hardships are just different from what traditional warriors experience. Experiencing the killing of another human being together is the greatest hardship endured by the crew. Boredom is a close second. Stress from constantly working and switching shifts is yet another. The pressure from authorities is another. And, finally, belonging to a community that is misunderstood and marginalized both in and out of the military is another hardship bonding RPA crews together.

Anonymity: The Unknown Killer

Perhaps the greatest contributing factor to group absolution in RPA strikes is anonymity. For the same reason an old-time executioner wore a mask, humans want to remain anonymous when engaged in the dirty business of killing. It's counter to our instinct to survive as humans to deny someone else that same opportunity. But when we can remain anonymous, that changes the dynamic.

Aggressive anonymous behavior is evident in how people behave on social media when their identity is concealed. In fact, some of the most aggressive individuals you will ever meet are online trolls. Those same individuals would probably behave like harmless sheep when face-to-face. The difference is anonymity.

The term *deindividuation*, which describes the effect of remaining anonymous in a group, was first used by the psychologist Leon Festinger in the 1950s to explain how individuals in groups conform to the group's behavior and lose accountability for their personal actions when they think they are anonymous. G. I. Wilson in "The Psychology of Killer Drones—action against our foes; reaction affecting us" thinks that RPA crews are affected by this phenomenon of "deindividuation":

> The anonymity in turn feeds into the commission of a violent act or acts that people otherwise would not normally participate in or be a willing accessory to . . .
>
> A significant number of studies have demonstrated that individuals who believed their identity was unknown were more likely to behave in an aggressive and punitive manner.

Anonymity does exist for RPA crews. It's a very secretive community in which information about operations and the people conducting them is well guarded. They fight as a group, diffusing responsibility among themselves and to others within the "kill chain," including JTACs, ground force commanders, or the Combined Air Operations Center. Proof of this concept is the use of call signs. Call signs are in no way unique to RPA operations, but they are the primary means of anonymity within the group.

Military units use call signs when communicating over radios. Call signs are either nicknames or an alphanumeric string of numbers and letters that identify a unit. Call signs help for brevity, operational security, and to distinguish which unit you are talking to on the radio. But call signs also mask your identity, providing anonymity. Call signs help strip away personal identities, both for the aircraft they fly and for the individual crew members.

Even the Combined Air Operations Center, a large operations center where many people work, uses a call sign, which actually gives it the trait of personalizing this vast organization down to a single entity. But while it personalizes the organization, it also obfuscates where a particular decision originated within the organization. It diffuses responsibility.

Military aviators are sometimes given personal call signs as well, often as a rite of passage. Personal call signs, such as Maverick, Goose, or Iceman in *Top Gun*, are used to communicate with one another during a mission. Personal call signs also help the crew adhere to the practice of "no rank in the box," which means that everyone on the crew is an equal contributing member and anyone can speak up regarding concerns about the mission or its safety. It enables an honest conversation to be held among a crew outside the normal rigid rank structure and military chain of command, and it facilitates operations. It's much harder to tell whom you are talking to if someone calls and says "this is Bender" as opposed to saying "this is Lieutenant Colonel Phelps." You can hide behind a call sign; it's like a verbal mask for your identity. Interestingly, there may be a benefit to using call signs for crew members to manage the dual identities many crew members described having earlier.

Conditioned to Kill Within a Group

The conditioning to work as a group begins on the first day of military service and continues throughout one's period of service. The military goes to great lengths to take away a person's self-identity in the beginning of boot camp or basic training, to make them realize that survival depends on being part of a group (i.e., deindividuation). Service members have similar haircuts or hairstyles, wear the same drab uniform, and relinquish the given name they have been called their entire life only to have it replaced with a rank and last name, or just a last name.

The military trains its members to do everything as a group, such as formation runs, close-order drill, employing a weapon as a crew, eating, sleeping, and in some instances even showering. The effect is to break down an individual through stressful situations, and to build back up as a group that depends on one another for survival. That's why during basic training the group gets punished by the drill sergeant/drill instructor for the "crimes" of an individual. It reinforces that the group is the purpose of being and the individual who joined the military no longer exists. John Keegan discusses this in *The Face of Battle*:

> The wearing of uniforms, however variegated, however splendid, dimin-
> ished the individual identity of combatants, which it has been one of the
> principle functions of the medieval panoply to emphasize. So too did the
> imposition of a rigid chain of command, which robbed subordinates of
> that independence by which the headstrong nobleman had always set such
> store, while the new insistence on drill reduced the individual soldier's
> status to that of a mechanical unit in the order of battle.

This conditioning from day one prepares service members to kill in most instances as a group and it is by design. (This was in the fine print of the contract in case you didn't notice.) RPA training is no different, and in many ways might be the best example of this concept. The training of RPA crews (including munitions training that is simulated, inert, or live) is almost an exact replica of the real GCS. With minor differences, the training is practically seamless to real sorties and weapon employments as compared to a fighter pilot who is in a simulator, not the actual jet.

Even if all that conditioning does not take hold, studies have shown that people tend to behave for the roles they are assigned or the role that is expected of them. A classic study of this behavior is the Stanford Prison Experiment led by the Stanford psychology professor Philip Zimbardo in 1971.

The experiment took place on Stanford's campus in a mock prison built in the psychology department's basement. Twenty-four volunteer students were assigned roles as either a guard or a prisoner by a random coin flip. The volunteer prisoners were initially "arrested" by the Palo Alto police and brought to the mock prison, where they were stripped naked, deloused, and issued a prison uniform (in some cases a dress to emasculate the male prisoners). They were also forced to wear a pantyhose-style cap (to simulate shaving their head, which occurs at most prisons) and issued a prisoner ID number to replace their name.

All these impositions served to dehumanize the participating "prisoner," shock their senses, and enforce conformity to prison rules; they are all similar to techniques used in military basic training to break down the individual. The volunteer students serving as guards were issued no guidance other than to do what they thought was necessary to run a prison and keep the prisoners in line. So the guards organized their own set of rules. When the prisoners failed to conform to those rules and tried to revolt, the guards exerted their authority through physical and psychological punishment.

For six days, they behaved in these roles just as they thought real prisoners and prison guards would. By the sixth day, the experiment was stopped by Dr. Zimbardo because it had gone too far, with instances of abuse of power by the guards and some emotional breakdowns by some prisoners. This kind of experiment would not be conducted today for ethical reasons, but we know that the behavior exhibited in Zimbardo's experiment does happen in real life, as it did in the Iraqi prisoner abuse at Abu Ghraib prison in 2004.

Like most facets of RPA warfare, it isn't that simple to say that deindividuation occurs among the crew or that they experience a mob mentality that enables them to kill as a group during the mission. Some of these group dynamic factors are certainly present and contributive: the crew remains anonymous from its victims, diffuses responsibility, holds one another accountable, and experiences hardships together. That, coupled with the distance from which killing occurs, should make an RPA the ultimate crew-served weapon. But there is a plethora of factors that actually work counter to deindividuation within RPA crews.

First off, the crew's actions are highly scrutinized, both within and out-side the military. Mistakes often make national and world news. Because of this scrutiny, there is a level of accountability that doesn't exist with other crew-served weapons. Counter to anonymity within the group is the fact that an RPA crew's actions are captured, documented, recorded, and widely broadcast to various locations. The crew's actions are observed by many authority figures, both within their chain of command and beyond it. There is no hiding what the group is doing and that fact alone denies anonymity.

For similar reasons, it has become standard practice for some law enforcement officers to wear a camera while on duty. It holds them accountable for their actions, and also protects them with documented evidence of their decisions and actions. When the camera is rolling, there is no dark shadow where a mob mentality could take hold.

Quite the opposite, actually. In the same way, all the people coordinating with the RPA crew for a strike are ensuring that the strike is conducted legally while eliminating or mitigating collateral damage. The long endurance of remotely piloted aircraft enables tactical patience, effectively defusing the violent contagions that occur within a mob. There is no rush to violence.

RPA crew members are susceptible to the factors that contribute to group absolution: anonymity, shared hardships, diffusion of responsibility, and accountability to peers. When combined with the fact that the RPA crew is never in close physical proximity to those they kill, it should be the ultimate crew-served weapon. However, the manner in which RPA are employed, observed, and scrutinized serves to reduce the absolution of the group, ultimately having an effect—both positive and negative—on the response to killing remotely.

Chapter Eight

Target Attractiveness

In Afghanistan I worked in the RCSW [Regional Command South-west] general's COC [Combat Operations Center] and remember the first time I witnessed a strike. Two men were hit with a Hellfire missile. One adult male was in pieces and the other was placing the pieces of his friend into a wheelbarrow, eventually giving up in his futile attempt and making a run for his life.

My work had some effect on these operations, but I did not control these directly. I just had the video feed up on the wall like it was a movie night on the watch floor. At first when I saw this, I felt a little sad or sick, even. I thought, *What kind of a person gets pleasure and enjoyment out of seeing one man explode into pieces?* Then I realized these men were in the process of embedding an IED in a dangerous area where U.S. Marines had lost their lives in IED explosions. Marines had been carried out of that area. Once they were whole, and now they were going home legless or limbless altogether.

I then started feeling a bit of pleasure in knowing we got one of them in a manner fitting to that of which they got us. Like an eye for an eye. I also had satisfaction in knowing that this one guy was never going to blow another limb off an American again.

—From personal correspondence with a United States Marine sergeant,
intelligence analyst

Attractive Kills

RPA crews spend countless hours watching people. Inevitably, when your job consists of watching people going about their daily lives you are going to witness some horrific things and some absolutely weird things. It's the nature of the business. An RPA crew may know more about some targets than they know about their best friend, while other targets will remain relatively obscure. Some targets will appear as normal as one's neighbor, while others will assume a more traditional appearance as an enemy.

Due to the long endurance of RPAs, within a single mission a crew may shift among several targets with unknown variables in relatively short order. What results is differing degrees of knowledge of, or context for, the people RPA crews are killing. In other words, understanding the "why" behind the targeting of an individual being watched so intently can vary greatly. I refer to this "why" as target attractiveness.

The crew's knowledge, context, and (to a degree) understanding of the target is immensely relevant to the ease or difficulty of killing, as well as to an individual's response to the kill afterward. Similar to the relationship between the resistance to killing and distance is the relationship between resistance to killing and target attractiveness. Awareness that the target has engaged in significant nefarious activity makes killing easier to rationalize. Figure 6 depicts this relationship between resistance to killing and target attractiveness. Simply put, the more attractive the target, the less resistance to killing on the part of the RPA crew.

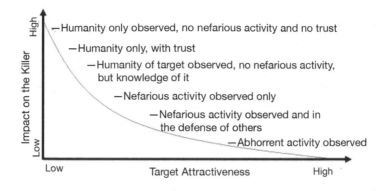

Figure 6

On the far left of the diagram, the most difficult kill to rationalize in the mind of an RPA crew is when they observe the humanity of a target and have no context or knowledge of anything else about the target and have no trust in those commanding the strike. By humanity, I mean someone going about their everyday life, conducting normal human activities such as going to the market, meeting friends for tea, or playing with their children. In this instance, the crew witnesses no nefarious activity or hostile actions committed by the target, they have no context or history of any nefarious activities committed by this individual, and they may not have complete trust in those authorizing the strike of the target about its validity.

While the target has been deemed legitimate by those requesting the attack, this situation produces the most resistance to killing. There is a seed of doubt as to why this individual is being targeted in the first place, and the crew may question the overall legitimacy of the strike.

Trust: Extremely Important When Killing at the Request of Others

Why would a situation even exist where the crew was questioning the legitimacy of a strike? Several reasons. The most common: an RPA may be dynamically retasked to support a ground force commander who has all the context of the target but may not have the time to build trust with a striking aircraft by providing that context. The requesting unit just needs the target prosecuted rapidly. Or the context of the target may exist but it can't be shared with an RPA crew in a timely manner prior to a strike due to classification or communication issues.

Trust between striker and those requesting a strike is vitally important for the targeting process to succeed. In the NPR article "The Warfare May Be Remote, But the Trauma Is Real," an Air Force staff sergeant working as an intelligence analyst told of how significant context about the target was for her to perform her job:

> In the early days of the fight against ISIS, [Staff Sergeant] Alicia says she struggled with some of the work she was being asked to do. "We were striking a lot at that time, and for me, it felt like I wasn't getting enough of the story behind the strikes," she said. "I like to know everything I can about what we're gonna be doing, and for me, that was what was bothering me. I felt like I didn't know enough."

An Army MQ-1C Gray Eagle pilot told us a story about how important trust with the unit requesting a strike was to him. He and his crew were flying the aircraft back to their base after a mission when they transited through an area where a JTAC was requesting immediate assistance from any aircraft to prosecute a target. Since the Gray Eagle crew had sufficient fuel and was closest to the JTAC, they were dynamically tasked to support him. Upon establishing contact on the radio, the Gray Eagle crew was given a building to target, with the explanation that it was full of enemy fighters. The trust in the requester was assumed and Gray Eagle and the JTAC worked together to put a Hellfire missile into the building. In the aftermath of the strike, the Gray Eagle pilot observed a man running out of the building holding a little girl in his arms. The man continued to his car, then sped off down the road with the girl. The JTAC requested an immediate reattack, this time on the car.

At this point, the dynamics of target attractiveness had changed, and the trust had eroded. Why the JTAC requested this second strike was unknown to the Gray Eagle crew. The context was never communicated. It could have been that the target was deemed so important to kill that the authorizing entity was willing to accept collateral damage. There may not have been time to share this information with the RPA crew. Or it simply could have been a mistake. In any case, the Gray Eagle crew refused to strike the car, and told the JTAC they were low on fuel and had to return to base. The JTAC was angry because the target wasn't struck again. An enemy fighter had been allowed to escape death to fight another day.

But the story doesn't end there. To demonstrate the amount of pressure that RPA crews are under to strike targets, this particular crew was brought up for discipline charges by a general officer in charge of operations in the area for refusing to do their duty. The Gray Eagle pilot even received a phone call directly from this general asking him "who he thought he was to deny a request for a strike from a ground force." The charges were eventually dropped when the full details of the situation became known, but this anecdote serves as a prime example of the external influence and pressure levied on RPA crews to "just strike the target."

When that trust *has* been established between those requesting a strike and the RPA crew, we see less resistance to killing. This is partly because of a diffusion of responsibility, but related to the crew's ability to rationalize—to process the act and accept it. In short, there is less cognitive dissonance and therefore less of a mental mismatch between what the crew is being told (or not being told, in some instances) and what they believe. Leveraged by the level of trust with the strike authority, this allows the crew to mentally invest in the action equally and allows rationalization to occur with less resistance.

Sometimes, the Context of the Target Matters

If the crew witnesses only the humanity of the target but does know the context of prior villainous activity committed by this individual, the target becomes more attractive and the resistance to kill lessens. In this circumstance, the context of why the individual is being targeted is known and trust with those requesting the strike exists already.

As much as the rationalization process depends on trust, the context relating to the target's attractiveness can matter almost equally. But which targets are the most attractive, and is there a target more attractive than others? Most RPA aircrews would agree that any target whose killing prevents a friendly force (to include coalition and partners) from being harmed or put in harm's way is the most attractive and easiest to rationalize. Hell, that's probably every soldier's most attractive target. Protecting friendly forces isn't a situation that exists in every RPA strike, though. What then amounts to the most attractive target outside protection of friendlies is high-value individuals.

By nomenclature alone it would follow that HVI targets are the most attractive. Their value is both assessed and tied to their military gain at the strategic, operational, and tactical levels. However, in some cases, these HVIs did something, or are responsible for something, so heinous that, in a sense, it outweighs their military gain; an act so egregious any soldier would willfully execute the deed when the opportunity presents, even without trust from the authority figure.

There is a justice aspect in striking this type of HVI. And unfortunately for this class of targets, the trust is already established with the strike authority because the terrible acts are known. This combination, outside of friendly force protection, makes this target type the most attractive and presents little to no resistance to killing. And, if the history of RPA warfare is any indicator, it does not matter how long ago or how many degrees of separation—targets in this class of bad guys don't get to hide. One such example is the strike on Al Qaeda commander Jamal al-Badawi in Yemen in January 2019.

In 2004 al-Badawi was convicted of plotting and taking part in the bombing of the USS *Cole* on October 12, 2000, where seventeen U.S. sailors perished. Subsequently, in 2004, al-Badawi was imprisoned in Yemen but then released for reasons unknown to the U.S. The U.S. military spent the next fifteen years tracking him down and on January 1, 2019, he was killed in an RPA strike while driving alone in his car in Yemen.

From here we move into areas where the hostile acts or sinister activities are observed by the crew—for instance, planting an IED on the side of the road or driving around with an obvious military weapon such as a technical vehicle (which usually consists of a pickup truck with a heavy machine gun or other weapon mounted in the vehicle). While no other context or knowledge of the target may exist, this is a fairly easy situation for the crew to rationalize. They are striking in defense of future others. This is definitely an attractive target and one most likely struck without much hesitation.

Striking a target in defense of others may become even easier. Responding to a request for assistance when troops on the ground are under fire is a situation that requires no rationalization other than protecting friendly troops. While an RPA crew may not have any additional knowledge of the target, nothing else is needed to protect their fellow warriors on the ground. To be clear, this doesn't mean that the humanity of the target is not observed in these situations, it means simply that the humanity doesn't overshadow the hostile act. One such example was provided by an Air Force Reaper sensor operator who supported Army Special Operations Forces (SOF) in January 2016 in the deadliest place on Earth for the U.S. military at the time, Nangarhar Province in Afghanistan:

It was not often that our assets were shifted to provide air support during troops in contact, but today American SOF were in the fight and we answered their call. We arrived late to the party as the SOF troops on the ground had killed several combatants. I changed my camera to infrared to search the tree lines surrounding the friendlies and after careful inspection my pilot and I located a small group of eight ISIS fighters engaging in small arms fire toward the friendlies. We immediately set strike posture and received our clearance to fire two Hellfires at the combatants.

I swapped my sensor to Day TV prior to weapons impact. The result: four enemies eliminated and one enemy wounded. The next thirty seconds showed me the enemy is not always vile, that they're actually human. As the pilot maneuvered back to strike posture and I maintained eyes on the still squirming wounded enemy, one of the remaining three fighters ran from cover, carrying his rifle and wearing a white scarf, threw his comrade over his shoulder, making a beeline for the trees.

We rifled another Hellfire and meters before he made it to cover, he and his comrade exploded under the missile I guided in. We eliminated the remaining fighters with our final Hellfire but throughout the entire engagement what stuck with me was the humanity, the heroism, the man wearing the white scarf showed in the face of death. All of our men on the ground made it safely to their beds that night, but I will never forget the man in the white scarf.

Lastly, any target caught in the act committing an abhorrent act such as an execution, beheading, murder of children, or rape requires no additional context or rationalization. The evil act is enough to elicit strong emotions affirming the belief that this target must be eliminated. In fact, when RPA crews are unable to strike these targets for any reason, there exists a sense of anger and oftentimes a feeling of failure that justice has escaped their clutches. This is made evident in the story one Air Force Reaper pilot relayed:

> After watching five Western hostages for over a month in a country we supposedly weren't in, I was told the rescue mission had been canceled because of political issues. We lost our legal authority to strike, so instead I had to watch them get beheaded one by one in front of a camera and couldn't do anything about it.

Observing Horrific Things

While this rationalization process occurs as depicted in Figure 4, it doesn't mean that an RPA crew is completely unaffected by situations that are easier to rationalize. In some instances, it is the nefarious act itself that elicits such a strong response for the crew. According to the 2017 NPR article "The Warfare May Be Remote, But the Trauma Is Real": "One Air Force survey found that among [intelligence] analysts engaged in this kind of work, nearly one in five had witnessed a rape within the past year. Some airmen reported witnessing more than 100 incidents of rape or torture." Other than first responders, there may be no equivalent occupational comparison to the carnage, evil, and horrors that these warriors witness. It's best just to let them describe these horrific situations in their own raw words:

- "On a few occasions, I've had to watch an execution and not been able to do anything about it due to either collateral damage concerns or not having enough information on the situation to discern enemy combatants." —USAF Captain, MQ-1/9 Pilot
- "The worst day was watching a Humvee full of Marines hit an IED and being unable to help." —USMC Staff Sergeant, RQ-2 Pilot
- "There was a time when I watched an Iraqi Army outpost get overrun by ISIS and the soldiers were killed." —U.S. Army Sergeant, MQ-1C Pilot
- "I watched the Blackwater civilians be strung up and burned in Fallujah." —USMC Staff Sergeant, RQ-2 Pilot
- "I watched ISIS kill multiple children for playing on a playground that was built by coalition forces." —U.S. Army Special Operations Sergeant, MQ-1C Pilot
- "I observed friendlies on the ground have a suicide bomber run up to them and detonate himself in front of them." —U.S. Army Specialist, MQ-1C Pilot
- "During deployment we had heavy weather rolling in as well as friendly troops in contact, which we were trying to provide overwatch for on a raid, but with the cloud decks and our limited visibility we weren't able to provide assistance and had to return to base. Come to find out, fifteen friendlies died as we noticed the bodies getting taken off a helicopter on the base where we were co-located with these friendly troops." —USMC Corporal, RQ-7B Pilot

- "I witnessed a man get buried by rubble from an air strike. He was only injured and did not die. But his comrades had to come and help dig his pinned body out of the rubble. I noticed him when only his right leg was exposed from the rubble kicking frantically, as if he was suffocating. I accidentally allowed myself to imagine being in that situation. I couldn't get that image out of my head after that moment." —USMC Sergeant, Intelligence Analyst
- "Watching the son of the person I just obliterated with a Hellfire missile pick up the pieces of his father. It wasn't the act of killing that I focus on, it was watching the boy's face and interactions with the rest of his family that continue to haunt me." —USAF E-6, MQ-9 Sensor Operator
- "Watching a young child carrying half of an adult male down a road to get help." —USAF E-7, MQ-9 Sensor Operator

The Unknown: Narrowly Avoiding Disaster

Some of the most intense emotions for RPA crews occur in situations where nothing actually went wrong, but disaster was narrowly avoided due to sheer happenstance, luck, or no action because something just didn't feel right. RPA crews are professional warriors who do not take killing lightly and they are always striving for the perfect shot. But this is war, and the "fog of war," as Clausewitz referred to it, is always present. A degree of uncertainty in combat almost always exists regardless of how hard we try to eliminate it. Dr. Peter Lee in *Reaper Force* refers to these situations as the "what ifs." I refer to them simply as the "unknowns."

One method used to mitigate the possibility that last-minute unknowns will surface during the RPA targeting process is what is called "shifting cold." Due to a missile's time of flight between launch and impact, the situation around the target can change in an instant. When this happens, the missile can be shifted cold to a safe, planned location away from the target where, in theory, no one will be harmed by the impact. This procedure is planned for in every strike prior to weapons release. While it works most of the time, it's not infallible.

An Army sergeant first class Gray Eagle pilot told us of a time when he was instructing a new pilot during a live mission. The aircrew had lined up to engage a building full of enemy fighters with a Hellfire missile, and at the last second women and children appeared in the vicinity of the target with the missile in flight. The Gray Eagle crew shifted cold into a wall adjacent to the building. Unfortunately, the Hellfire never made it to the wall, instead impacting a vehicle adjacent to the wall and instantly destroying it. While no one was harmed, in a poor country like Afghanistan, the Gray Eagle crew had inadvertently destroyed the vehicle, one of the family's most prized possessions.

Unknowns exist. One Reaper sensor operator told us of a mission in Afghanistan where he was following an enemy technical vehicle loaded with a recoilless rifle in the back. This rifle was a legitimate military target under the rules of engagement and, if allowed to be employed, could be very dangerous to friendly forces. He followed the technical as it traveled down a road for forty-five minutes, waiting for the moment when the vehicle would not be adjacent to other traffic—the right time to strike.

But the opportunity never presented itself. As the technical vehicle moved closer to an Afghan Army checkpoint on the road, the sensor operator was left with a moral dilemma: strike the target knowing that there would be collateral damage or let it approach the Afghan Army checkpoint, where coalition partners would most likely die. Thinking quickly, he passed the information about the vehicle's location and direction of travel over a tactical internet chat program and his radios. The information was relayed to the Afghan Army checkpoint, which was able to intercept the technical vehicle without incident. When the Afghan Army pulled the passengers out of the vehicle, among them was a ten-year-old boy. Disaster had been averted, but the sensor operator was left to deal with the knowledge that he had almost killed a child that day.

This is perhaps the worst nightmare of an RPA crew, and one that warriors will replay in their head over and over. How did I miss that detail? How could it have been prevented? How do I prevent it from happening again in the future? These unknowns are thoroughly reviewed post-mission, in an ever-evolving pursuit of perfection in anticipation of preventing an unknown from turning into a future tragedy.

Weird Things Observed as a Professional Voyeur

Most RPA crews are initially unprepared for the plethora of weird things they observe during their missions. A sensor operator told us that his one piece of advice for new people to the community is to "prepare to see some weird shit." He went on to describe the first time he witnessed bestiality, which lasted for more than an hour. While you might think this an unusual event, most people I interviewed had observed a man making sweet love to an animal.

Another sensor operator told of watching a man walk down to a riverbank to masturbate. Unbeknownst to the man at the time, it was his last sexual encounter. The crew killed him later that day. These acts are incredibly intimate moments in a person's life. For a professional voyeur, observing the performance of sexual acts is just part of the job, particularly in the parts of the world where people sleep on their rooftop to avoid the heat and things that go bump in the night.

And then there are the times when RPA crews stare at or look for inanimate objects for a really long time. One pilot told of staring at clouds for seven hours a day for a week straight hoping to catch a glimpse of enemy fighters below when the clouds dissipated. Another pilot was asked to look for a ten-foot-tall radio antenna in a pine forest. And then there's the ever favorite completely useless task of looking for signs of disturbed earth or hot spots or to report back anything unusual. In a land where sex with animals occurs so frequently that it's considered "usual," you have to wonder what would be considered unusual.

Dark Humor: A Survival Technique

Thus we begin to understand just how unpredictable, dark, and volatile every target can be, and the wide variety of circumstances and experiences in which an RPA crew member might be asked to observe and possibly kill. If nothing else, it shows killing with an RPA is much more than a binary categorization of yes or no, whether to pull the trigger. It is complex and speaks to the professionalism, expertise, and humanity of those at the controls. And the rationalization process in every scenario can be completely different based on the above factors. So how does someone cope with such mental yoyoing? It's simple: dark humor.

Dark humor in the military is certainly a survival mechanism. Warriors understand this odd practice. Sometimes you have to find the humor in an otherwise completely miserable situation to keep from going crazy. Well-timed inappropriate humor can relieve tension. But the humor can be understood only by someone who has also endured an equivalent miserable experience. It becomes a moment of truly "embracing the suck," laughing about it, and moving past it.

When this dark humor is discussed outside the situation or with others who have never experienced such a situation, it comes across as heartless, unprofessional, or immature. But laughter truly is the best medicine and dark humor is how the warrior administers it. Laughter says, "This has no power over me." Here are just a few of the favorite sick and twisted anecdotes relayed to us by RPA crews.

- We had been following a guy on a horse for weeks. He went into the mountains. We ended up shooting the horse and the female pilot got a call sign of "Elmer" for it, as in Elmer's glue.
- We were watching a circle of fifty extreme Taliban beheading lesser devout Taliban fighters. They were all enemy fighters, but we ended up debating at what point we should strike them. Do we wait until they kill all of the less extreme fighters and then strike when they have done half our work

for us? That seemed the best course of action so that's what we did.

- I saw four missiles impact near simultaneously on a large group of Taliban fighters doing calisthenics. They were so uncoordinated it was hard to tell who was wounded and who was still doing jumping jacks.

- There is a legendary shot within the community known as the "Angry Bird" shot. Just as in the video game, when the missile impacted the ground the enemy fighter flew fifteen feet into the air.

- A missile that turned out to be a dud actually impaled the guy we were targeting. It was a one in a million shot.

- I used to make up conversations for the people I was watching like it was a sitcom.

    Husband: "Honey, I'm home!"

    Wife: "How was your day? How many IEDs did you plant? Did you kill any infidels today?"

    Husband: "Always with the work questions as soon as I come in the door."

- During the Battle of Ramadi, Iraq in 2016, our Reaper was tasked to find and fix enemy forces at the front line. We observed a number of enemy forces moving tactically and called in their positions. Eventually we were cleared to engage one group that was moving toward friendly forces. We engaged and killed two enemy fighters at a street corner next to a wall. While remaining eyes on the impact site for battle damage assessment we observed two different enemy fighters heading toward the missile impact site from different directions. They could not see each other until they met at the corner, at which point one of the enemy fighters raised his rifle and shot dead the other enemy fighter coming from the other direction. We witnessed the fog of war firsthand and the subsequent realization that the combatant had killed one of his own.

In a very simplistic yet truthful sense, you can contrast the scenarios these RPA crews experience with those of an infantryman on the front line. Most likely the infantryman is being shot at by an enemy combatant. For this infantryman there is little resistance to killing, and it usually turns out that the person firing at him just became the most attractive target. The primary motivation is pure self-preservation. There's an obvious need to kill.

With RPAs, given the lack of physical danger, there is not an immediate "yes" button to justify killing (except for supporting ground forces). Thus, in the absence of being shot at and knowing who the shooter is in that exact moment, an RPA crew needs to trust the authority providing that information.

As we have seen time and again, context matters. It matters for the legitimacy of an RPA crew striking a target in the first place. It matters for the rationalization process as the crew deals with the almost unthinkable acts it must perform in carrying out its duties. And it matters, most of all, to help the crew members wake up the next day and watch horrific and weird acts all over again.

SECTION IV

Barriers, Help, and the Future

Chapter One

The Video Game Comparison

Shall we play a game?
> —Joshua (a.k.a. WOPR, or War Operation Plan Response) in the
> movie *War Games*

Video Games: Origins and Overestimations

Video games are a relatively new invention within human history and we are still gaining an understanding of the impact they have on society as they evolve. In October 1958, the physicist William Higinbotham developed the first video game built solely for entertainment. Higinbotham worked for Brookhaven National Laboratory's instrumentation group, and every year in October Brookhaven opened its doors to visitors for a tour of the laboratory to showcase its work. In an effort to create a more engaging experience for the guests, Higinbotham developed a simple analog video game called *Tennis for Two*, which displayed very basic graphics that permitted a player to hit a ball over a net.

Higinbotham thought that it would "liven up the place to have a game that people could play, and which would convey the message that our scientific endeavors have relevance for society." He was right. The video game was the hit attraction that year, with people waiting in very long lines for their chance to play video tennis.

In true physicist fashion, by October of the next year Higinbotham had improved the video game to allow guests to play video tennis on the surface of the moon or Jupiter, with the associated equivalent gravity and characteristics of how a tennis ball would react on the surface of those celestial bodies. The video display was also enlarged for a better guest experience. The game was once again a hit, but then it disappeared, failing to make a reappearance at future open houses.

No patent or invention followed since Higinbotham worked for a government lab and wouldn't have owned the rights to the invention anyway. It wasn't until 1964 that the first patent for a video game was granted to Sanders Associates, a defense firm that later sold the patent to Magnavox, which began developing video games for entertainment in the early 1970s. By 2020, video games were a $159 billion industry.

On August 10, 2018, a twenty-nine-year-old employee at Seattle–Tacoma Airport stole an empty seventy-six-passenger propeller airplane from a maintenance area at the airport and took it for an unauthorized flight over Puget Sound. There was no explanation of his motive for stealing the airplane even though he was in constant contact with air traffic control throughout the flight. The employee understood the serious nature of his actions and even stated that "this is probably jail time for life, huh? I would hope it is for a guy like me."

The local air traffic controllers tried to get him back on the ground safely. In the course of their conversation with him, the aspiring pilot made several odd statements over the radio. He asked for the coordinates of an orca whale carrying her dead calf on her back—a story that dominated the news at the time. He desired to go look at the Olympic Mountains. He expressed his concern over being "roughed up" or shot down if he complied with air traffic control and landed the airplane at a military base. He even asked whether Alaskan Airlines would give him a job if he landed the plane safely. Finally, he asked if he could do a barrel roll before "calling it a night." What was deeply concerning was his claim that he could land the airplane by himself because he had "played some video games." Tragically, the flight ended in a crash and the man's death.

The immersive and seemingly realistic experience of playing video games is becoming increasingly lifelike while simultaneously our relationship with the world around us is becoming more digitally interactive. These two phenomena are dangerously converging in an area where a person's ability to distinguish reality from virtual reality can be confused, as we saw with the airline employee at Seattle–Tacoma Airport and have seen with mass shootings whose perpetrators train using first-person shooter games.

When was the last time you tried to use your finger to swipe or alter a screen's display that wasn't actually a tactile screen? We guarantee it's happened to you at some point if you use a tablet, smartphone, or smart device. We are being conditioned to interact with our world as if we are playing a video game while simultaneously video games are conditioning us to be more violent.

We are also living in a time when the overestimation of one's abilities due to having "played some video games" is widespread. We errantly assume that because a video game is so lifelike it is a suitable replacement for proper training to accomplish such difficult tasks as landing an airplane. We also errantly assume that when humans interact with an object by manipulating buttons to control it in a fashion similar to playing a video game (such as an RPA's ground control station), the overall act is the equivalent of playing a video game. Both of these are slippery slopes that lead to dangerously false assumptions.

Why RPAs Are Compared to Video Games

Upon meeting an RPA crew member for the first time, people often say, "It's just like playing a video game, right?" Let's just get one thing clear: it's offensive and downright disrespectful to our warriors to compare the responsibility of taking a life to playing a video game. RPA crews may not always be so blunt when asked about the comparison, but that is what they are thinking. The comparison of video games to RPAs originates from a place of ignorance and is perpetuated by those who don't understand the professionalism and serious nature of the work being conducted by a military RPA crew. Nothing about killing someone from seven thousand miles away should ever be considered a game, and anyone who makes that assumption trivializes the serious nature of the business and marginalizes the person performing it. This misconception that our RPA warriors are just playing a video game is also a barrier to taking their psychological struggles seriously. Let's explore how this common misconception came into existence.

There are four main reasons that flying an RPA is equated to playing a video game: how RPAs are flown, how RPAs look, the level of training of the crew, and the remote nature of RPA employment. The comparison to video games arose from the similarities between how people interacted with early RPAs and computers.

One of the earliest RPAs employed by the U.S. military was the RQ-2 Pioneer. Inside the ground control station of the Pioneer, the heading, altitude, and airspeed were controlled through a series of twistable knobs that the pilot adjusted to reflect the desired flight attitude and profile. To take off and land the Pioneer, an external pilot used a handheld remote-control station with a joystick similar to that used to control a model airplane. The twistable knobs that controlled how the Pioneer was flown varied dramatically from the stick-and-rudder skills employed by a pilot in a traditional airplane. Instead, the Pioneer resembled early video games with joysticks. Over time, those flight-control functions migrated from the clunky manipulation of a knob to the push of a button on a computer mouse or the clicking of a slider bar on a computer screen, which further resembled video games.

An autonomous system that enabled takeoffs and landings with the push of a single button replaced the external pilot's use of a handheld remote-control station. Autonomy helped reduce human error over time, particularly in mundane tasks that were necessary for the execution of the flight but weren't overly important for the execution of the mission.

Early RPA systems were flown in a manner that resembled a model aircraft, and subsequent models were flown in a manner that resembled a computer game. So how early RPAs were flown contributed significantly to the video game comparison.

Early RPAs resembled model airplanes and were much smaller than the lethal RPAs employed today. The RQ-2 Pioneer, for example, had a wingspan of only seventeen feet and weighed 452 pounds. The Pioneer looked like a smaller unmanned version of an OV-10 Bronco, which had a wingspan of forty feet and weighed almost seven thousand pounds. It wasn't until the introduction of the RQ-4 Global Hawk and MQ-1 Predator that RPAs started to reach a size equivalent to traditional aircraft. Since the word *drone* encompasses a description of unmanned systems ranging in size from small hobby drones all the way up to lethal RPAs, it follows that they have all been lumped into the "toy airplane" comparison by the uninformed.

Before the U.S. Air Force became involved with RPAs in large numbers and started assigning officers to fly RPAs, early RPA systems were flown by enlisted pilots in the Army, Navy, and Marine Corps. Enlisted pilots were not rated aviators at the time and received varying degrees of aviation training based on their service, but nothing equivalent to their manned counterparts. The locations where initial RPA aircrew training occurred also were not the historical locations where military aviators trained, such as Naval Air Station Pensacola for Naval and Marine aviators. The training occurred outside of the "cradles of aviation" facilities and involved enlisted ranks, who hadn't served as pilots since World War II. These differences between manned aviation and unmanned aviation, with respect to both where the training occurred and who was receiving the training, were enough to encourage traditional aviators to downplay the skills and training required to fly an RPA. These important distinctions, combined with the fact that the cockpit never left the ground, made the RPA community an easy target of mockery and marginalization among traditional aviators. The predominant method was to equate the RPA job to something as simple as playing a video game, thus ensuring that traditional pilots retained primacy.

Another reason RPAs are compared to video games is the remote nature in which they are employed. The RPA aircrew isn't normally in the same physical space as the combat environment they are fighting in. Their actions through the RPA are observed on a screen, which can feel a bit surreal. Video games are obviously controlled through a video game controller, processed through the gaming system, and displayed on a screen. RPAs are controlled in a similar manner, processed through the ground control station, and displayed on a screen. It is not an apples to apples comparison, but it definitely appears similar enough that people feel comfortable making the comparison.

Because of the way RPAs are flown and the remote distance involved, there is a misperception that RPA aircrews' actions occur isolated from reality and that they become disassociated killers. John Govern in "The Importance of Distance of Modern Warfare" stated:

The most human elements of warfare become absent for those utilizing remote weapons systems. The moral dilemma of killing becomes distinctly diminished as human lives become blinking dots on a screen, extinguished forever by the touch of a button.

Laurie Calhoun in "The End of Military Virtue" stated: "Training [RPA aircrews] to kill in the manner of sociopaths with no feelings whatsoever for their victims because they are but icons on computer screens is a frightening prospect." A leading advocate for RPA in the U.S. Air Force, Colonel Joseph Campo, described this problem well in "Distance in War: The Experience of MQ-1 and MQ-9 Aircrew":

RPAs often distills into a two-pointed proposal regarding video-game warfare. The first point revolves around violent video games and their ability to desensitize people to horrible acts, violence, and killing. The second point states killing via RPA operations has effectively turned war into a video game for the RPA aircrew. These two points are often fused to paint a picture of video-game-playing RPA aircrew who have no understanding of the actual destruction their weapons are causing.

Additionally, even if RPA aircrew did understand the physical destruction their weapons caused, their upbringing and the technical nature of RPA operations has desensitized the aircrew to the point that they are unable to generate any true emotions or understanding of their actions.

As previously discussed, I wholeheartedly reject these assertions that due to the distance involved RPA aircrew lack emotion, connection to the mission, and an understanding of their actions. As evidence of RPA aircrews' awareness of their actions, one simply has to listen to their own statements. In "The Theory of the Drone 10: Killing at a Distance," one Predator pilot stated:

> We understand that the lives we see in the screens are as real as our own...I would not compare what I do as a job comparable to Call of Duty/any other video game, in any sense. It is very real and the seriousness of the lives on the ground is very real and instilled in all of our training. It is never something that we joke about. Very serious business.

In the same article, another pilot rebuffed the notion that he was dissociated from the mission:

> I wonder why people think this. We understand what we are doing is real world operations. We know our actions have consequences. I don't understand the idea of being desensitized due to some operators not being in an actual firefight/combat zone.

A further rejection of this misaligned theory was discussed by Air Force Reaper pilot Major Dave Blair and the psychologist Karen House in "Avengers in Wrath":

> [There can be a] zone of moral abstraction—a Tomahawk launching on map grids or a high-altitude bomber dropping through a Norden bombsight, wherein the potential immediate trauma of this kill is lessened by its lack of intimacy or personal knowledge of the target. One has

to deliberately think about the consequences of his or her lethal act in order to turn that abstraction into a visceral sense of death. This zone is presumed by many to be the "video-game-war" space occupied by remote warriors. We reject this description as applied to anyone in the military, and it is certainly not the space inhabited by the high-resolution sniper-like attack profile of Predator and Reaper crews.

The problem is that the video game comparison is made based on a perception of what people think an RPA aircrew does, as opposed to what they actually do. It's oversimplified and doesn't encapsulate the skills required for the job. Adding to the problem is the fact that the RPA community is a very tight-lipped, secretive community. Accurate information about the nature of the job is difficult to come by for outsiders, and access to observe the environment in which operations are performed is difficult to attain.

In America's wars in Iraq and Afghanistan, reporters were often embedded in military units, giving firsthand accounts of the struggles and challenges experienced by those warriors. Due to the secretive nature of the RPA community, reporters rarely embedded within the RPA community. A few who did had an anti-drone agenda and were looking for ways to reinforce their own opinion. What has filled this void of firsthand observation is inaccurate conjecture about what it's like to fly an RPA, reinforced through erroneous portrayals in movies and the media.

The misimpression about RPAs exists even within the military, where some wrongly perpetuate the video game comparison. Master Sergeant Madhur Sawhney, an RPA pilot in the Marine Corps, told us the story of a Marine sergeant major (the senior enlisted rank) who sat in his ground control station during a combat mission he flew. After ten minutes of observing everything the crew was doing, the sergeant major apologized to them for thinking that it would be like a video game and expressed surprise about the level of detail and professionalism required to do the job. Anyone who has spent time in a GCS during a combat mission, as that sergeant major learned that day, quickly realized that flying an RPA is no game.

RPA Similarities to Video Games

Are there similarities to flying an RPA and playing a video game? Absolutely, or the comparison wouldn't exist in the first place. Due to the remote nature of employment, RPA aircrew and video gamers both remain physically safe from harm, regardless of how stressful and hectic the situation appears on the screen. In both cases, actions in the real or virtual world are enabled by the push of buttons and the results are displayed on a screen.

Another similarity is how some smaller RPAs are flown. What really isn't helpful to keeping a professional distinction between reality and video games is that some smaller RPAs use PlayStation or Xbox controllers as their human–machine interface to manipulate flight controls or the sensor. This choice of controller device perpetuates the stereotype that RPAs are just like video games. Ironically, this controller device was chosen because a large portion of aircrew were already conditioned to use a video game controller, so it shortened the time required to train them. It's a common tool.

One Reaper sensor operator I interviewed said that first-person shooter games prepared him very well for an RPA cockpit when it came to the multitasking skills required to be successful. He made sure to state that the comparison ended there. Another Reaper sensor operator told me that there is a bit of relevance to the statement that flying an RPA is like playing a video game. The problem, as he put it, is when it becomes too easy to kill. He chose not to expand, but the point is taken that when separated by a distance and mediated through a screen, without the possibility of harm to oneself, an RPA crew might kill with less hesitation than if face-to-face with the enemy.

Finally, and perhaps most important, killing is rewarded in both instances. In a violent video game, killing helps the gamer accrue points, defeat competitors, and advance to higher levels. Violent video game behavior is rewarded, which in turn conditions the gamer to engage in future violent behavior. Killing with RPAs is rewarded by saving fellow service members' lives, by achieving a goal that has been worked on for a long time (such as killing an HVI), or by taking actions that are just and help defeat one's enemies, contributing to the overall military or national objective.

RPA Differences from Video Games

How is playing a video game different from flying an RPA? The main difference is obviously that in a video game the killing isn't real. That should be obvious on the most basic level. For some reason, however, this simple fact doesn't prevent people from comparing a highly trained military RPA crew to a teenage boy playing *Call of Duty* in his mom's basement. RPA aircrew understandably loathe this comparison and were very vocal when I asked them their thoughts about it during interviews. I chose to let the aircrew describe their reactions to the comparison:

- "Video games don't show a wife and kids' anguish after you 'shwack' their dad and watch them pick up his body parts. The immediate aftermath is the worst." —USAF MQ-9 Sensor Operator
- "A video game comparison makes it sound like we don't understand what these actions entail. The other half is that we don't do a good job about talking about it. When there is a lack of information, people's biases take over." —USAF Captain, MQ-9 Pilot and former C-130J Pilot
- "It doesn't make me angry to hear it. I've been hearing it for thirteen years. I'm desensitized to the backlash of comments against noncombat arms. No video game prepares you for the decision-making that occurs. It's not that hard to fly a robot, but the cerebral functions are drastically different (from a video game)." —USMC Staff Sergeant Christopher Herr, RQ-2/7/21 Pilot
- "In one, somebody dies. In a video game, no one dies." —Former military intelligence analyst who works for a three-letter agency's RPA program

- "It's frustrating as hell. It's utterly professional. It's the cockpit of an airplane that stays on the ground. It detracts from the difficulty of the motions of what we have to do. It's more about communication than it is about the act of flying the airplane." —British Military Intelligence Coordinator for MQ-9
- "There could be nothing further from the truth. There is joy in a video game. The amount of expertise required is beyond any sort of video game. Video games aren't scary because there are no real ramifications." —USAF MQ-9 Pilot
- "It doesn't even cross my mind as a good analogy, as a good comparison. A video game doesn't carry the weight or responsibility of fratricide, collateral damage, loss of life of friendlies. To me it's overly simplistic. The average person isn't familiar with what the cockpit of a UAS even looks like." —U.S. Army CWO3 Eric Cooper, MQ-1C Gray Eagle
- "It's nothing like a video game. There may be some similarities of stress between completing a campaign. Within two seconds I can kill a person from seven thousand miles away. It's the exact same amount of pressure and the stress is the same as a pilot in the cockpit." —USAF MQ-9 Sensor Operator
- "I hate that comparison. People don't really understand what we do. That is their way of trivializing what we do." —Master Sergeant Madhur Sawhney, USMC
- "If you were to compare this to a video game, this would be the hardest job you would ever have to do with the worst training to do it. There is still a risk to human life no matter what. They don't have a concept of what we do as a community of warfighters." —USAF MQ-9 Sensor Operator

And from Campo's article "Distance in War: The Experience of MQ-1 and MQ-9 Aircrew":

MQ-1/9 aircrew were also queried on whether they considered RPA operations to be a video game and how did they feel about such a comparison. In response, aircrew were unanimous in their statements that RPA operations are not a video game. This point cannot be overstated. Every interview participant, regardless of whether they were an 18xer (RPA pilot), previously aircrew, or experienced a positive or negative psychological response to killing, were united in stating that RPA combat operations are not akin to video gaming.

Yet another difference was described by Denise Chow in "RPA Wars: Pilots Reveal Debilitating Stress Beyond Virtual Battlefield":

Critics say firing weapons from behind a computer screen, while safely sitting thousands of miles away, could desensitize pilots to the act of killing. What separates this, they argue, from a battlefield video game? In video games, players rarely make a human connection with the characters on their screen, but Predator RPA operators often monitor their targets for weeks or months before ever firing a weapon.

Does Playing Video Games Make You a Better RPA Pilot?

The United States Air Force is obviously sensitive to the video game comparison and has made some efforts through public affairs and marketing to set the record straight. One such communication was a webpage post by Technical Sergeant Nadine Barclay from the Public Affairs office of the MQ-9 Wing. The article, titled "Busted, Top 10 RPA Myths Debunked," addressed head-on the misperception that RPA pilots are just "gamers":

Fact: Our Airmen are trained to be the best pilots in the world, regardless of aircraft. Our fully qualified aircrews consistently exceed expectations for both flight safety and operational effectiveness. Like pilots in manned aircraft, RPA pilots are required to meet the same qualifications. New RPA pilots undergo a very intense training program before they fly operational missions. This training curriculum lasts approximately one year, and many current Air Force RPA pilots and trainers have already completed undergraduate pilot training in manned aircraft as well.

Unfortunately, that statement reads like an oversimplified government institutional response to an actual complex situation. "We train them well so it can't be a video game..." What if we trained them well on a video game? What if I told you that the U.S. Air Force is actually considering incorporating more simulation and virtual reality training into its future undergraduate pilot training for manned aviators? This training would look more like a video flight simulator or a video game than actual time spent in a traditional airplane. The Air Force calls it "Pilot Training Next" and it may just be the solution to solve their pilot shortage: reducing the time to train new pilots in a more cost-effective and efficient manner.

Pilot Training Next is described as cutting-edge technology that blends advanced biometrics (such as wearable devices to monitor heart rate, breathing rate, pulse, and stress levels), artificial intelligence, and virtual reality. It will be self-paced, and student pilots will even be allowed to practice flying with their own virtual reality headset at home.

What the Air Force has found in early tests of Pilot Training Next is that students experience more in-depth learning, learn faster, and are exposed in a controlled environment to difficult situations that sharpen their airmanship skills. In the first test run of Pilot Training Next in 2018, thirteen of twenty student pilots completed the undergraduate pilot training in just four months. Normal undergraduate pilot training takes a year to complete. It may not be a real airplane, but it's real enough to condition the mind into learning how to fly one.

Even if playing a video game conditions you to be a better RPA pilot or a traditional pilot, that still doesn't equate a combat mission to a video game. It's not the mechanics of how an RPA is flown that is important. It's the serious nature of the job and the consequences of actions that matter.

To the uninitiated and uninformed, RPA warfare may appear like a video game, with "desktop warriors" dissociated from combat pushing a button and killing without remorse. Nothing could be further from the truth. This prevalent sentiment, though, is a major obstacle to admitting that RPA warriors can suffer psychological trauma from seven thousand miles away, and this false narrative of the video game comparison is a barrier that can prevent those warriors from getting the help and support they need from their chain of command, from the mental health community, and from the public.

Chapter Two

Culture

As the longbow overcame the advantages enjoyed by armored knights and steam abruptly ended the quiet solitude of sailing to meet one's enemy on the high-seas, so too does the advent of weaponized RPAs represent a risk to the current hierarchy among warriors and the military bureaucracies that administer them. . . .

Years later we may come to realize that our strong convictions about warfare and weaponry were superseded long ago, but we were blinded by emotion or bureaucracy and failed to recognize the change occurring all around us.

—Colonel Joseph Campo, "Distance in War: The Experience of MQ-1 and MQ-9 Aircrew"

Culture: What the Hell Is It?

Why is military culture important in a discussion about killing with RPA? A poor culture can lead to many barriers for the warriors within it. Ultimately a poor culture can lead to the false notion that within a community issues do not exist, problems may be less likely to be reported, or people may be discouraged from seeking help when needed. You do not have to look any further for proof than the example of how our Vietnam veterans were treated. For those who do not understand what we did to our Vietnam veterans, the reader is encouraged to read the section called "Killing in Vietnam: What Have We Done to Our Soldiers" in *On Killing*. A negative culture can also lead to poor job satisfaction, which in turn feeds into burnout, cynicism, apathy, stress, and diminished attention to detail.

Culture has as many definitions as specimens, but it is most often defined as the customs, arts, social institutions, and achievements of a particular nation, people, or other social group. Culture has also been defined as the outlook, attitude, values, morals, goals, and customs shared by a society. For the sake of characterizing the culture of a group within the military, we will define *culture* as what a group values, how a group views itself, how others view the group, and how the group is treated.

There are numerous subcultures within the military, generally aligned to their military occupational specialty (MOS); in layman's terms, their job. If you imagine each military subculture as a zoo animal, the RPA community would be a flying platypus, unique and misunderstood in almost all aspects.

In a biblical analogy, the RPA community is Eve, formed from the rib of manned aviation. Specifically meaning that the RPA community has a borrowed culture that has been thrust upon it, and one that may not be the best fit for how it operates, what it values, and how it views itself. But there has been little room for a creative discord that would make it possible to acknowledge and assert that the RPA community deserves its own culture.

It is safe to say that for nearly two decades the RPA community has been trying to conform to a manned aviation culture in which it doesn't quite fit. That has been the standard defined for it, what has been modeled, what is familiar to aviation leadership, and what is familiar to the manned aviators doing a tour in RPA units. Similarly, in America's efforts at "nation building," we build things in our own image, things we are familiar and comfortable with, regardless of whether it's right for that culture. The RPA community has been no different, born from manned aviation and formed in its likeness.

Wings and Things

An important aspect of cultures are ceremonial artifacts, jewelry, and clothing. These are some of the most outwardly visible and obvious indicators of a culture. Within an aviation culture, aviators wear flight suits and earn aviation wings that are pinned on their uniform to identify them as aircrew.

Thus, our RPA warriors wear flight suits and earn their wings during flight training. Is it necessary? Does it serve a functional purpose? It is really hard to justify the functional purpose of a flight suit for a ground-based RPA crew. Flight suits are made of flame-resistant material, which is important when riding in a giant hunk of metal full of highly flammable aviation fuel, but not important when the cockpit is nowhere near the fuel and never leaves the ground. No, the reason RPA crews wear flight suits is purely cultural: to appear like their big brother, manned aviators. It's not easy to justify to other cultures in the military why RPA personnel wear a flight suit. While this is less of a cultural issue in the Air Force, in the Marine Corps it is a constant source of friction, criticism, and defense. The U.S. Army's RPA crews don't wear flight suits, which is further proof that it is a cultural issue rather than a functional one.

While the flight suit may serve no purpose for RPA personnel other than to look the part of an aviator, the wearing of wings is important because it indicates a level of training completed, skill achieved, and professionalism expected. Wings put unmanned aviators in a similar class as manned aviators. However, the wings are not identical to manned aviators', nor are they separate but equal. The difference in appearance between manned and unmanned wings is subtle to an outsider but glaring to someone within the aviation community. Those glaring differences tell a story, oftentimes of a cultural oppression of the unmanned community.

As one example of a slight: The Marine Corps approved the wearing of UAS wings for its pilots and sensor operators in 2018, nearly thirty years after the formation of its first UAS unit! The initial design of the insignia had a single anchor in the center, golden wings on the sides, and a shield in the middle with a V-like symbol emblazoned on it (the tactical symbol for UAS). The single anchor was important. Naval aviators, better known as pilots, have a similar set of wings: single anchor (to denote a pilot and express naval roots and traditions), golden wings, and a shield in the center with no symbol. The same wing design with two crossed anchors represents a Naval flight officer (NFO), someone trained to do a specific job, other than flying the aircraft, as part of the aircrew. It is an aspect of the Marine and Naval aviation pecking order, subtle but real: pilot first, then NFO. When the Marine UAS wings came back reluctantly approved by the institution, the single anchor for the UAS pilots had been replaced with two anchors.

There would be no mistaking an unmanned pilot for a manned pilot within Marine aviation. The pecking order was maintained. And when opposition was raised to the final approved design of the wings, the community was told to "just be happy that you even got wings," which is the equivalent of a frustrated parent telling a child "because I said so." These actions tell unmanned aviators, "We placated you with some wings, but we will never consider you a 'real' pilot." Hopefully by now it is apparent that the flying platypus needs a culture of its own, instead of living in the shadow of manned aviation.

A Fun House Mirror: How RPA Crews View Themselves

Ask five RPA crew members how they view the community and you will get ten different opinions in response. Generally speaking, RPA personnel view themselves as professionals who are proud of the work they do. They communicate clearly that they are having a large impact on combat operations, and they argue that their support to the soldier on the ground is unparalleled by any other aviation asset. In my survey, nearly half (46 percent) said they were very satisfied with their job. Only 9.5 percent said they were either very or somewhat dissatisfied. The remainder of the results in the chart below depict a community with a large majority of satisfied warriors. This is a positive and surprising sign considering the myriad issues we have discussed within the community and the grueling nature of the occupation. It may well be that the vocal minority is gaining the most attention while the silent majority just professionally soldiers on.

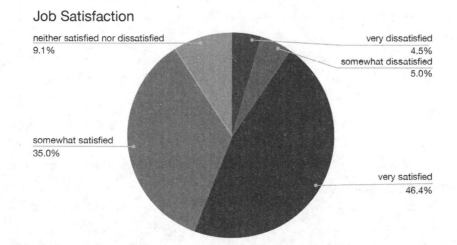

Job Satisfaction

neither satisfied nor dissatisfied
9.1%

very dissatisfied
4.5%

somewhat dissatisfied
5.0%

somewhat satisfied
35.0%

very satisfied
46.4%

Despite the self-reported satisfaction, we know that RPA personnel face a plethora of cultural issues. RPA pilots spoke of an operational tempo so taxing that they are not afforded the same opportunities as other officers, such as attending the career-level schools needed for promotion. These pilots aren't asking for special treatment, just an equal opportunity, but they are keenly aware of and outspoken about this perceived dis-enfranchisement. Some aircrew spoke of a frustrating mixed message from their branch of service: on one hand there is an insatiable appetite for RPAs while on the other there is a tendency to shun or mock those associated with RPAs.

One Reaper pilot told us that he didn't even feel as though he were part of a community, much less part of the Air Force. He worked on an isolated base surrounded only by other RPA members in an odd microcosm. This isolation, coupled with his constantly changing shifts and the secretive nature of his job, made him feel as though he were missing out on the larger experience of comradery and esprit de corps typical of the service.

In a time when the millennial generation seeks jobs in which they can have an impact as an individual, it's easy to see why, despite the numerous issues encountered in the RPA community, job satisfaction remains high. Perhaps Cherie Armour and Jana Ross summed it up best in their 2017 article in *Military Psychology*, "The Health and Well-being of Military Drone Operators and Intelligence Analysts: A Systematic Review":

> This finding suggests that despite being emotionally exhausted with a sense of indifference towards what they are doing, the majority of drone and intelligence exploitation operators believe that their work has value and they are making a significant contribution in their role at work.

Tired but satisfied, doing meaningful work—this accurately reflects the community sentiment.

Cultural Fratricide: How RPAs Are Viewed Within the Military

As soon as Orville Wright landed after the first flight at Kitty Hawk, he and Wilbur probably headed straight to the local pub for a six-hour debrief. Following the debrief they undoubtedly decided they needed a way to announce to the world that they had courageously mastered a machine that defied death and defeated gravity and were now "officially better than everyone else." Or at least that's how I imagine it went down between the first two aviators.

After Kitty Hawk, aviation rapidly evolved into a culture of daredevils, risk-takers, and popular-culture heroes, particularly in the areas of death-defying aerial acrobatics, dangerous endurance flights, and aerial dog-fights pitting skill against skill to the death. And so the aviation culture continued for one hundred years, producing warriors who performed magnificently, with the ultimate litmus test being courage to overcome fear of flying in life-threatening situations. Which is why the explosion of RPAs and the corresponding lack of physical courage required to operate them is met with cultural chagrin from manned aviators. To manned aviators, RPAs represent an erosion of their cultural values.

There is bound to be friction when a culture that respects physical courage above all effectuates a subculture that doesn't require physical courage to fly. This friction exists for two main reasons: RPAs disrupt the traditional model of physical courage expected of a pilot, and RPAs threaten to replace manned aircraft in certain capacities. Manned aviators have the most to lose, so they fight back with the aggressive spirit for which aviators are renowned.

What would you call someone who graduates from flight training and flies an aircraft? *Pilot* would be the logical term, unless that aircraft is an RPA. Then it gets confusing; but the job title is important because it indicates status within the culture. Instead of using the existing and accepted aviation job descriptions of pilot or weapon systems officer (Navy and Marine Corps) or combat systems officer (Air Force) to define RPA crew positions, the job titles are on the periphery and out of synch with aviation—i.e., sensor operator, air vehicle operator, mission payload operator, unmanned aircraft commander, and UAS officer. The closest job title to *pilot* is RPA pilot in the Air Force, but the preceding qualifier of "RPA" ensures there is no confusion that a distinction with a "real" pilot exists. Roger Connor from the Smithsonian National Air and Space Museum touches on this point in "The Predator, a Drone That Transformed Military Combat":

> Another challenge is cultural—who is a pilot? Initially, most RPAS pilots for the U.S. Air Force were experienced combat pilots, but demand soon exceeded supply and the military services began training non-pilot

operators. This has created organizational frictions in the military over who has the privileges of pilot status in a world where unmanned and autonomous operations are increasingly important.

Grégoire Chamayou in *Drone Theory* captured a commonly held opinion of RPA personnel expressed by another pilot:

I simply scoff at the idea of some computer nerds whining about "battle fatigue" or "PTSD" when they not only know what they are getting into but aren't even in the same country getting shot at. It's a slap in the face to those who *really* deploy, who *really* get shot at and who *really* have to deal with the psychological effects of war.

Remember, the aviation culture values courage. Lack of an opportunity to display physical courage in the performance of the assigned job makes you a second-class citizen in the eyes of those taking risks. Pratap Chatterjee in "A Chilling New Post-traumatic Stress Disorder: Why Drone Pilots Are Quitting in Record Numbers" reaffirms that "the pilots themselves say that it's humiliating to be scorned by their Air Force colleagues as second-class citizens." This sentiment is pervasive throughout the culture. I offer one anecdote from my own time as a commanding officer of a UAS squadron working adjacent to an attack helicopter squadron known as Scarface.

More Kills Than Scarface

Within the military community of warfighters there is a pecking order. In the Air Force fighter pilots reign at the top while in the Marine Corps the institution is built around the infantry. Even within subcultures of the larger service, such as aviation, there is a pecking order of "meat eaters." That pecking order is traditionally defined by the individual's initial knowledge and skill needed to enter the community, and afterward by the destruction that their platform can bring to bear on the enemy.

Initially it starts during basic flight training. The pilots with the best ability, scores, and aptitude tend to migrate to fighter and strike aircraft or attack helicopters. This isn't to say that transport helicopters or mobility aircraft aren't filled with quality pilots, but the more aggressive "type A" personalities usually end up in shooting platforms. RPAs are unique in that they are disrupting that tradition.

As a commanding officer of a Marine Unmanned Aerial Vehicle Squadron, my squadron was part of a larger Marine Air Group (MAG) consisting of transport helicopters, attack helicopters, and tilt-rotor aircraft. There was always tension between the attack helicopter squadron and us. The attack helicopter squadron, full of UH-1Ys (Hueys) and AH-1Ws (Cobras), fancied themselves the meat eaters of the MAG, the steely-eyed killers. But during my time as the commanding officer, my squadron had elements deployed for twenty-two months hunting down insurgents and facilitating their death, while Scarface remained at home. We were damn good at it, too, with more than one hundred confirmed enemy fighters killed as a result of our support.

Whenever a pilot from Scarface would give us a hard time for flying a remote control "toy" or playing a video game, I'd simply brush it off and state that we had "more kills than Scarface this year." It irked them to know that we were contributing to the fight while they remained at home in a training status. It bruised their ego, but it was indicative of the larger problem that the RPA community faces every day within the military. RPAs are disrupting the traditional model of who is delivering violence to the enemy.

This is especially true within the Air Force, where RPAs have been employed with great success and in numbers comparable to their manned counterparts in recent conflicts. The meat eaters are being dethroned and they don't like it. Culturally, RPAs threaten the virility of manned aviation and in some circumstances its very existence. But the focus needs to be on the best weapon for the job, and increasingly that weapon no longer includes a pilot in an airplane. For the first one hundred years of aviation it did, though, and that is a lot of history, culture, and resistance to overcome. It is quite possible that RPAs are the most effective killing aircraft ever built that are not considered cool to fly by other pilots. In a culture of meat eaters and risk-takers, killing in a cool way also defines status.

Nobody Puts Baby in the Corner: Promotions, Medals, and Opportunities

The way we treat a community of people conditions their response and their behavior. It also demonstrates how much the community is valued by others. Operant conditioning is accomplished through either positive or negative reinforcement or positive or negative punishment. Reinforcement aims to increase a behavior, whereas punishment seeks to decrease a behavior. The methods used to change the behavior are either positive, adding something such as a reward, or negative, taking away something such as a privilege. In positive reinforcement, the subject is rewarded for an action. This creates a feeling of wanting to perform the same action again for another reward. In the military, positive reinforcement is rewarded through promotions and medals.

As we know from *On Killing*, "This process of associating reward with a particular kind of behavior is the foundation of most successful animal training." One would think that a community that works this hard and has an overwhelmingly decisive impact on combat operations would be commensurately positively reinforced.

The promotion rates for RPA pilots lagged behind all other pilots in the Air Force from inception until 2013. The U.S. Senate asked the Government Accountability Office (GAO) to look into this issue in 2014 and again in 2018. The GAO found that promotion rates for RPA pilots did indeed lag behind all other pilots in the Air Force from 2010 to 2013. From 2014 to 2015 the rate of promotion surpassed only that of mobility pilots, and from 2016 to 2017 it returned to the lowest promotion rate. The rate significantly increased in 2017. Promotion rate data for 2018 was not included in the study.

One reason for lower rates, as admitted by the Air Force in the study, was that in the early years of RPAs lesser-performing manned pilots were transferred to fill the demand in the RPA community. The Air Force attributes the increase in promotion rates for RPA pilots after 2014 to the fact that it created the 18X RPA pilot job and now the ranks are filled with a similar quality spread of officers as other pilot categories. Thus far, this positive reinforcement of promotion for performance for RPA pilots has been insufficient. In a positive sign, the GAO concluded that by 2018 opportunities for promotion, education, and non-flying tours for RPA personnel had, after lagging behind for so many years, increased to rates similar to those for other pilots.

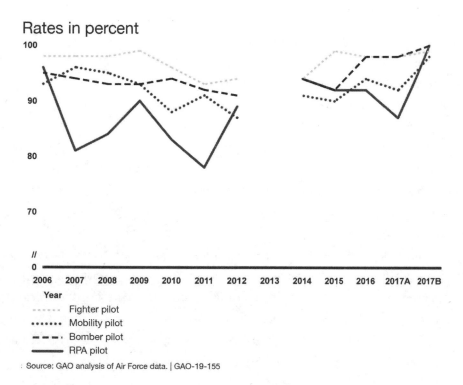

Rates in percent

Year

- - - - - - Fighter pilot
••••••• Mobility pilot
— — — Bomber pilot
———— RPA pilot

Source: GAO analysis of Air Force data. | GAO-19-155

Promotion Rates from Captain to Major for Remotely Piloted Aircraft (RPA) Pilots Compared with Pilots in Other Career Fields from 2006 through 2017, Except 2013.

Concerning the second opportunity for positive reinforcement, awards, we find one of the biggest medal controversies of recent times. In February 2013 Secretary of Defense Leon Panetta announced the creation of a Distinguished Warfare medal for those occupations that have a significant impact on operations but didn't fit the traditional requirements for medals based on heroism or performance in combat. The Distinguished Warfare medal was designed for RPA and other military occupations such as cyber warfare that are critical to success on today's modern battlefield. The problem was not the creation of the award itself, but the fact that it was a higher award than a Bronze Star.

The Distinguished Warfare medal was the first newly proposed combat medal in seventy years. The outcry over the precedence of the award, which quickly came to be dubbed the "Nintendo medal," was brutal and relentless and came from veterans communities, active service members, and the media. Justifiably so. It was a horrible idea to place the Distinguished Warfare medal above a medal that required risk to life and courage on the battlefield. The Distinguished Warfare medal's proposed position in the military awards' order of precedence didn't do the RPA community any favors and unnecessarily put them in the crosshairs of outrage from nearly everyone. A month after it was proposed, Secretary of Defense Chuck Hagel, who had replaced Panetta, canceled it with instructions to search for another way to honor the contributions of RPAs and Cyber Warfare. What emerged in 2016 was a well-thought-out distinguishing "R" (for remote combat operations) that attached to existing personal medals.

Another interesting quagmire surrounding awards and RPAs relates to campaign medals. In order to rate a campaign medal, a service member must be physically present in a combat zone for a designated period of time. It is therefore entirely possible for an RPA crew to kill more enemy combatants in a campaign than those who are physically present fighting in the country, and yet still not rate an award that delineates their participation in that operation. At the same time, someone could deploy to a combat zone, never leave a forward operating base and never engage or even see the enemy and still proudly wear a campaign medal. The awards system within the DoD is historically slow to evolve, so it is no surprise that the department is lagging behind in figuring out how to handle this situation.

Why are promotions and awards even being discussed in the context of how RPA personnel respond to killing? The authors of "Psychological Dimensions of Drone Warfare" summed it up extremely well:

> Some experimental research indicates that greater social invalidation of a killing act predicts greater feelings of distress. One Government Accountability Office (2014) report suggests drone operators are commonly given fewer opportunities for advancement or medals, given the perception that they are not equivalent to traditional combat pilots. To the extent that stress is not balanced with recognition or accolades, this may increase the likelihood of burnout or PTSD in operators.

Which is exactly what has happened.

How RPA Personnel Are Portrayed in the Media: Disengaged, Damaged, and Derogatorily

The portrayal of RPA personnel in the media is disgraceful, uninformed, and inaccurate, and further serves to marginalize the entire community. Most accounts depict RPA personnel as disengaged from the fight due to their physical distance, damaged due to the amount of killing they do, or overall derogatorily due to the political agenda of the critic.

Name one good movie or TV series in which RPAs are portrayed in a positive, favorable light. Amazon's *Jack Ryan* series portrays a conflicted dumbass who swaps money with his crew after a kill, goes to Syria to meet the child of a man he killed, and is an emotional dumpster fire from what he has seen and done. Andrew Niccol's movie *Good Kill* is melodramatic, hyperbolic, and full of inaccuracies, and is openly mocked within the RPA community as a horrible misrepresentation. *Good Kill* claims to be based on true events, widely assumed to be about former Air Force sensor operator Brandon Bryant, who was briefly consulted for the movie.

There is no doubt that Bryant, later diagnosed with PTSD, suffered as a result of his time working in RPAs. But his story is recycled in every anti-drone media account and incorrectly regarded as the standard for how the entire community reacts, responds, and handles the stress of the job. The media has taken a single data point and adopted it as the fact that supports their anti-drone agenda. Bryant himself expressed his disappointment about the movie in the 2015 *Newsweek* article "What 'Good Kill' Gets Wrong about Drone Warfare," stating, "I wanted [them] to make a powerful movie, not just an entertaining one . . . They're doing what our society does—marginalizing the traumatic effects of personal experiences."

The term *marginalize* in its original definition meant to write in the blank space (or margin) adjacent to the main text of a document. It meant to put something on the periphery of what was important. That definition became overshadowed by the metaphorical definition of relegating a person to an unimportant or powerless position. *Marginalization* is an accurate term for how the media regards RPA personnel.

Perhaps the greatest indication of media marginalization is the way RPA warriors are labeled. RPA personnel have shamefully been referred to in media accounts as "armchair" warriors, "desktop" warriors, "computer" warriors, "cubicle" warriors, "Nintendo" warriors, "PlayStation" warriors, and, a personal favorite, the "Chair Force." That last one is so funny it's hard to be offended. However, when the military writ large permits marginalization of a group of its warriors, those warriors start from a position of disadvantage as to how seriously they are viewed. And the taking of a human life is serious business. From now on, drop the pejorative descriptive in front of *warrior* to describe RPA personnel. A simple description of "warrior" will suffice.

Lower promotion rates, insufficient awards, lack of opportunities outside the community, a negative media portrayal, and a negative perception within the military are all symptoms of a poor culture resulting from insufficient advocacy. And advocacy starts with generals.

Advocacy: Fighting an Uphill Battle Without a General

There are general officers who advocate for the RPA community, but there are no RPA generals. The community is still too young. Even for those generals who do advocate on behalf of RPAs, it is often an uphill battle against the institution. Retired four-star General John Jumper described the struggle he faced in the early stages of just getting the Predator armed and into the fight. According to Jumper, a former Air Force chief of staff: "There is no other single enterprise in the Air Force that has produced results like [remotely piloted aircraft]. I had to kick down doors. If I had not been a four-star general, [arming the RQ-1] never would have happened." Advocacy at the general officer level is undoubtedly what it takes to effect change within the military.

Traditional pilots in the general officer ranks lead aviation within each branch of the U.S. military. They are the advocates who decide what is important for their service regarding funding, manning, opportunities, promotions, and in most cases procurement of aircraft. Essentially, they decide the plans, policies, procedures, and procurement: the future direction of aviation within each branch of service. They are the leaders who have the ability to significantly change a community or culture.

Lacking an RPA general with a seat at the decision-making table stifles the voice of the RPA community. It is extremely important for a community to have a general officer advocate who grew up in the community and understands its challenges, issues, and idiosyncrasies. The sooner we have such generals, the better. The great value of RPA to our nation's defense and our warfighting ability is irrefutable. The sustainability of this effort, and the long-term well-being of the RPA crew who are called by their nation to provide this essential, critical service, requires that general officers rise up from the RPA ranks. These general officers can then provide the priorities and guidance that will establish a healthy culture for this community.

Chapter Three

A Toolbox of Best Practices

I would just say there is one misperception of our veterans and that is they are somehow damaged goods. I don't buy it. There is also something called post traumatic growth....Not everyone reacts the same way, but I don't buy that somehow if you came home from Iwo Jima or Gettysburg or Iraq or Afghanistan, that somehow you're limited in what you can do. The Greatest Generation came home from WWII, the worst war in world history, and they created good communities, they rose to be college presidents, started industries...I just don't buy that somehow we are handicapped because we've been in those circumstances. I recognize grim realities. I don't recognize the limited potential of a human being when they come out of that.

—General James Mattis

Robots, Battleships, and an Invading Army

There is another famous story about RPAs that is worth telling. A story that actually sounds like the description of an action thriller movie involving the U.S. Navy, robots, battleships, and an invading army. And while it may sound like fiction, it is indeed fact. Forty-six years after the Japanese Instrument of Surrender for World War II was signed on the deck of the USS *Missouri*, remotely piloted aircraft were launched from the same battleship to search for Iraqi targets to destroy during Operation Desert Shield and Desert Storm.

Throughout the war, RQ-2 Pioneers from Fleet Composite Squadron Six (known as the VC-6 Firebees) were flown from both the battleships USS *Missouri* and USS *Wisconsin* to serve as forward observers for the ships' big 16" guns. The story of the RQ-2 employment is best told by the official command history of VC-6 for 1991:

VC-6 Pioneer Unmanned Air Vehicles (UAV) played a critical role in support of battleship combat operations throughout Operations Desert Shield and Desert Storm. VC-6 Detachment Patuxent River was formed into two sub-detachments during the Gulf War. Detachment ONE deployed with USS WISCONSIN (BB-64) and Detachment TWO deployed with USS MISSOURI (BB-63). Manned and supported at a level intended to support only one surveillance flight every other day, the detachments flew three to four flights daily during the war to provide real-time intelligence, target acquisition, identification, gunfire spotting, and real-time Battle Damage Assessment (BDA) which would have been impossible in the heavy AAA [antiaircraft artillery] environment with manned aircraft.

The UAV not only provided targeting information, but vital intelligence necessary to determine when to hold, cease or continue fire. The result was greater tactical flexibility and management of available ammunition. For example, when a single shot from USS MISSOURI hit an ammunition bunker and caused multiple secondary explosions, real-time BDA showed that no further rounds were required on that target. Another target was selected. In all, 1,224 rounds were directed at enemy positions in southeastern Kuwait with deadly effectiveness.

On its most crucial mission, a UAV detected a major Iraqi armor movement heading south toward U.S. Marine positions. The information was passed to USCENTCOM which diverted a formation of attack air-craft already in flight to destroy the enemy tanks. Other mission successes included a Silkworm missile site which was identified by a UAV as a decoy, allowing for critical U.S. assets to be directed elsewhere. An on-station UAV over Faylaka Island photographed Iraqi soldiers waving white flags, dramatically recording the first ever surrender of enemy forces to an unmanned vehicle. This imagery was one of the Navy's most memorable media events. Footage was shown on world-wide TV and frames were printed throughout the international press.

Dale Bohannon, the maintenance officer for VC-6 Detachment One aboard the USS *Wisconsin*, recalled to us the details of the first recorded surrender by an enemy to an RPA, which took place on March 1 and again on March 3, 1991:

There were a number of Iraqi troops on Faylaka Island. The bridge to the mainland had been bombed and they had been cut off long enough to be short of supplies. They had indicated their surrender and arrangements were made to remove them from the island. We were tasked to survey the island ahead of the [coalition] helicopters' arrival and fly overwatch for the evacuation of the Iraqi troops.

Here's where the story gets fuzzy. We had heard that the Iraqis knew that sound [of the RQ-2 Pioneer] usually meant that ordnance was soon on its way. We had shelled the dummy Silkworm site on the south end of the island several times. I assume the *Missouri* had as well. The fast movers probably saved one for it, too (someone did eventually get it). The reason we were there on the first and third was to look things over to ensure this wasn't a trap. Technically, they had already surrendered. That's why we were there. I've always assumed they wanted to make sure that we knew they had already surrendered.

By March 1, 1991, the Iraqi soldiers had associated the distinctive and very loud sound of the RQ-2 Pioneer with shelling of their positions. The Iraqi troops on Faylaka Island had already signaled their willingness to surrender, but on March 1, when they once again heard the loud buzz of the RQ-2 overhead, they were taking no chances and appeared holding anything they could fashion that would resemble a flag of surrender. News of the attempted surrender to the Pioneer reached General Norman Schwartzkopf, the commander of all allied troops. It was said, according to Mark Rindler, a member of Schwartzkopf's staff at the time: "When he [General Schwartzkopf] saw the surrender video of enemy troops to an RPV [remotely piloted vehicle] on Faylaka he was elated and said words to the effect that he believed this was the first time in warfare that enemy troops had surrendered to a robot." It is, in fact, known to be the first time that anyone has attempted to surrender to an RPA. Today, that intimidating RQ-2 Pioneer, aircraft #159, is on display in the Smithsonian National Air and Space Museum.

Pioneers during Desert Storm had an impact on the war and captured a pivotal moment in history. In the decades that followed, there were many more pivotal moments involving RPAs. It is apparent that use of this tool of warfare will only increase in the coming years. We owe it to the future generations of RPA warriors to prepare them for the unique struggles they will encounter. Throughout this book I have discussed possible issues of killing remotely and potential impacts on RPA crews. We need to change some of our procedures for the long-term health of the force and the sustainability of these operations. That is why, based on my observations, I offer up a toolbox of best practices for killing remotely.

Preparation: Interviewing Potential Killers and War Stories from the "Old Sarge"

The military excels at preparing its warriors for the act of killing but does a poor job of preparing warriors for dealing with the aftermath of the act of killing. It's partly by necessity. We cannot afford for our warriors to get bogged down in a pity party, mentally paralyzed by what their country has asked them to do during war. But there is room for improvement.

No problem resolves itself by festering beneath the surface and going unaddressed. This especially includes dealing with any negative responses to killing.

The solution within the RPA community is twofold: better screening in the initial entry-level process and more discussion during training with the "old sarge" than has gone before us. Some military occupational specialties exist that have only a slight possibility of being involved in combat actions against the enemy. This is not the case with the RPA community, especially armed RPAs. The probability of killing the enemy in times of conflict for armed RPA crews is near certainty. Therefore, we owe it to those entering the RPA community to ensure they are not only aware of this from the start but are well prepared for the possibility. Treat it like a job interview, where the job just happens to involve the taking of another human life, because it does.

The Air Force leads the services in this effort, including discussions during entry-level training about the possibilities of killing associated with the RPA job. The U.S. Army, which has armed RPAs, and the Marine Corps, which plans to have armed RPAs in the future, do not have these discussions during entry-level RPA training. Although no person really knows how they will respond to killing, an initial discussion during entry-level training will at least reinforce that it is part of the job. And that leads to the second point.

The people with the knowledge to guide and prepare the newbies through the experience of killing with RPAs are the "old sarges, old captains, and old chiefs" who have done it. The old sarges who have the experience of killing from an RPA owe it to those who are new to the community to discuss potential psychological and physiological responses. The old sarges know that during an engagement someone may experience an elevated heart rate, shaky hands, and time distortion. The old sarges know which missions will emotionally affect someone the most, because they have lived it. The old sarges can best prepare the next generation for the challenges associated with this unique environment because they have already navigated it. But those conversations between experienced warriors and newly minted warriors must occur at some point in the training to normalize the expectations and possible responses from the beginning. From *On Combat*:

> It is largely a twentieth-century affectation, a modern, self-inflicted psychic wound, to believe that you will be mentally destroyed or emotionally harmed by the act of killing during lawful combat. I am convinced, based on interviews with hundreds of men and women who have had to kill, that if you tell yourself that killing will be an earth-shattering, traumatic event, then it probably will be. But if you do the rationalization and acceptance ahead of time, if you prepare yourself and immerse yourself in the lore and spirit of mature warriors past and present, then the lawful, legitimate use of deadly force does not have to be a self-destructive or traumatic event.

RPAs are no different in this regard, and the experienced warriors in the community will prepare the next generation.

Breaking the Intimacy

As I discussed in the chapter on intimacy versus dehumanizing the target, RPA crews may form a one-way empathetic intimacy with those whom they are observing and targeting. This is in large part due to the long pattern of life missions that often precede striking. The intimacy can be further intensified when the warrior remains on station after the strike to survey the aftermath—in other words, watch the targeted individual's loved ones mourn his loss. It is the mental equivalent of raiding a building, shooting an enemy combatant, searching them, and finding intimate and humanizing possessions (such as pictures of their family). But, in the case of RPAs, the situation is unfolding in real time as our crews are watching spouses and children hysterically pick up body parts of the person just obliterated with a missile.

There are two possible ways to break this intimacy. The first, and most unlikely but possible scenario, is to have an aircraft that did not conduct the strike conduct the post-strike analysis. Logistically and tactically, this is impractical. But a second, more plausible possibility may serve a dual purpose: having a "swing crew" available to relieve the striking crew in the ground control station. Anytime a crew is working up for a strike, the swing crew could come to the GCS and build their awareness of the situation to prevent a loss of tactical continuity. Post-strike, the swing crew would conduct the battle damage assessment, while the striking crew leaves the GCS, gains their composure, takes a break, and discusses the event. In other words, breaking the link of observing the aftermath of the strike while simultaneously providing some time for the crew to mentally process what just occurred. This could well be the single greatest mechanism for mitigating the psychological trauma that impacts our RPA crews, and it is something that I highly recommend.

Shift Work: Eliminate the Rotating Shift

As discussed in the chapter on sleep and mental armor, sleep is vitally important for recharging our battery. A healthy sleep cycle is essential to maintaining a continuity of rest and recovery. Every disruption to the sleep cycle may affect the crew in a negative way. Rotating shifts are the most harmful. The medical research in this field tells us that the practice of rotating shifts can take years off a warrior's life, and it is deeply harmful to families. (Families can handle day shift or night shift, but rotating shifts appears to destroy them.) Rotating shifts gives everyone the equivalent of jet lag, thereby reducing the performance of RPA crews and increasing their vulnerability.

Over time, sleep deprivation can wear down the RPA crew's mental armor, allowing traumatic experiences to seep in and have a negative effect on them.

For a solution we once again turn to law enforcement officers, who have been tackling a similar challenge for well over a century. The best method appears to be a schedule based on a point system for seniority, in which you "bid" for your shift and earn the shift you want through longevity and seniority. We should also reduce or eliminate the rotating shifts for everyone because all they are accomplishing is creating a less effective fighting force. The mind and body will adjust to a shift over time, but frequently rotating that shift under the guise of "fairness" does no one any good. Eliminating rotating shifts would reduce operational stress, family issues, health issues, retention issues, and job satisfaction issues. Eliminating rotating shifts is the single most effective change that can positively influence the long-term health of the crews and the sustainability of operations.

Method of Employment: Less Remote and More Connected

The unprecedented remote nature of RPAs has certainly introduced some challenges never before experienced in the military, including the transitioning from combat to home on a daily basis, supporting combat operations without ever deploying outside the continental United States, and killing from seven thousand miles. These are byproducts of remote split operations in the U.S. Air Force and British Royal Air Force. The U.S. Army and U.S. Marine Corps still forward deploy RPAs. But the trend is moving toward more extreme remote operations among all the services.

Extreme remote employment may actually be contributing to feelings of surrealness associated with killing during missions. There is no frame of reference of combat operations for those who have never deployed and no distinguisher that serves as a demarcation line between combat and everyday life. We need our warriors to understand the culture, people, and environment that they fight in. The best way to develop this understanding is to deploy.

Using RPAs may be efficient tactically, logistically, and financially, but the cost is paid in other ways, as we have seen. There are three potential solutions: forward deploy the ground control stations and crews closer to the area of operations, relocate the ground control stations away from units' home station and deploy the crews for four to six months to support operations, or deploy a warrior to a launch and recovery element (LRE) prior to serving on the mission control element. At some point a warrior needs to deploy, and that point is early in the service member's career.

The first possibility is the least likely to occur. There exists a vast infra-structure that supports operations in this chosen method of employment and to alter it would bear significant cost.

The second option, however, may be plausible. The units support-ing operations could deploy away from their home station to support operations for a period of time, even if that location remained inside the U.S. The benefits would be tighter unit cohesion: training together for deployment as a unit building a peer support network, deploying together, and conducting a post-deployment transition prior to returning home. This would mimic the combat deployment of units that serve in a foreign country. It would eliminate the daily transitions between home and combat and provide a period of focus for the RPA warrior. Due to the disruption of the normal home life routine, there would be a clear distinction of combat operations and noncombat operations.

The last option of deploying warriors forward to serve on an LRE should be a mandatory first assignment for RPA crews whenever feasible. Every RPA crew member I spoke to who had deployed on an LRE said he had a better understanding and appreciation of the work he was doing stateside as a result of the deployment. Deploying more RPA personnel closer to combat may also provide a good frame of reference for the warrior and prevent the surreal feelings reported to occur during operations.

Debrief: Vital for Success and Health

The importance of a post-mission debriefing cannot be overstated. With the embedding of the Human Performance Teams (HPT)—teams made up of operational psychologists, chaplains, and medical professionals—RPA crews in the Air Force now have unfettered access to help when they need it. As the other services expand their RPA operations, they should consider providing a similar support team. While HPTs are a fantastic idea, embedding them within the unit may lead to unintentional barriers to warriors seeking help given the close proximity of peers. In other words, warriors may be less likely to seek help if their peers know they are doing so, despite the great efforts to destigmatize the support available. There are very few secrets in a military unit, and embedding an HPT may entail sacrifice of anonymity for those seeking help.

Some RPA warriors told us they felt that the help offered by HPTs is only lip service. The responsibility rests on the warrior to self-report to an HPT member, oftentimes in front of their peers. This may lead to lost opportunities for valuable discussions. Armour and Ross reinforced this observation in "The Health and Well-being of Military Drone Operators and Intelligence Analysts: A Systematic Review":

> Considering the need for close collaboration among the drone crew members, it is possible that the concerns about stigma are even higher in this occupational group (especially when the ask for help is done in front of your peers).
>
> Increased access to medical care was reported by a significant propor-tion of drone and intelligence exploitation operators, however, there were also reports of decreased utilization of health care services, primarily due to the difficulties of scheduling daytime appointments around one's shifts. Employing specially trained mental health professionals who understand the nature of the drone and intelligence operators' work as well as the nature of their working environment may help to tackle this problem and may also reduce the mental health stigma in this population.

For the events that are known to cause the most negative responses to killing, there should be a mandatory debriefing with the HPT at specific time intervals removed from the event. To begin the process, take away the requirement to self-report or self-identify. Discussions with the HPT should just become accepted after an event known to cause trauma, such as a strike with civilian casualties. The initial discussion should be followed by multiple debriefings with the HPT at various intervals to gauge the mental health of the warrior involved. This normalizes the process and reduces the stigma associated for those seeking help.

The composition of the HPT is well thought out, but it is missing a vital component: an experienced peer. One way to bridge this gap would be to include an old sarge within the HPT. The HPT has all the theoretical knowledge and education to help, but not the practical experience of killing with an RPA. The old sarge has the experience. Including a trusted peer could help get a struggling warrior to open up after a tough mission and is a tried and tested method that works.

Lastly, debriefings are vitally important for identifying the successes and failures of a mission. But is it absolutely necessary for the crew to watch the videos of their shots over and over again, searing the events into their memory? That does not seem productive for the long-term health of the crew. Which is not to say there isn't a benefit to watching the "game-day tapes," but might it be better to watch and analyze another crew's engagements as opposed to your own, and swap notes at the end? That would provide for objective input on performance while simultaneously breaking any intimate connection with the target or mission that may have existed.

A Break from Killing

Military units need a break from operations and from killing. Respite allows warriors to be removed from the stressful rigors of combat, refocus on training, take time off to spend with family or friends, attend military schools, and just rest and recharge prior to resuming high-tempo operations. Imagine that a palm tree blowing in the wind represents an RPA warrior. During normal operations or training, the tree is unaffected by the wind, but during combat operations the wind increases to hurricane-level strength. A tree can last in this wind for a limited duration before it snaps. A dwell period or respite allows the tree to straighten up and improve its chances of weathering future storms.

As discussed, this downtime or "dwell" period does not exist for Air

Force RPA units. Instead, Air Force RPA squadrons are in a perpetual state of supporting combat operations—a tree in constant hurricane-force winds. Blair and House in "Avengers in Wrath" describe the problem well:

> Even a squadron of warrior-philosophers, without adequate rest, would devolve to a *pro forma* approach to wrestling with the weight of taking lives. Given the endemic exhaustion in the Predator and Reaper community, we must ask ourselves if we are giving our remote warriors the ability to process the actions we are asking them to undertake. We have not (yet) structured our remote combat community around the recognition that remote combat is actually combat, and cannot be a permanent state of being. Therefore, we find crews in a perma-war footing, never less than 72 hours from the possibility of taking a life, for years at a time, with the exception of leave and TDYs [temporary additional duties].

Two possible solutions exist: create a deploy-to-dwell model similar to that in other branches of the military or reduce the weekly work tempo below fifty hours a week per person.

When a squadron is to have a dwell period from operations, another unit must fill the operational requirements. To accomplish this dwell goal requires either a reduction in the overall amount of RPA flights conducted worldwide or an addition to the number of RPA units. Neither option is easily executed or likely to happen.

If a lengthy dwell period cannot be carved out, perhaps a more sustainable model would be to reduce the amount of time that an RPA crew works each week—below fifty hours. In the studies conducted by Dr. Chappelle on Air Force RPA crews, those who worked more than fifty hours a week demonstrated more operational stressors and issues. A reduction in work hours to below fifty per week would require more aircrew on the line but would not require the addition of new units. Something to consider as we shape the force of the future.

Law enforcement agencies throughout the world have figured out how to run continuous operations in a demanding high-stress environment without breaking their people. For long-term sustainability of RPA operations, we should learn from this community and consider adopting a schedule similar to that used by law enforcement agencies.

Change the Culture for the Sake of Future Generations of RPA Warriors

A positive culture is important for the long-term viability and sustainability of the RPA community. The current culture desperately needs revamping. There should be an RPA culture that accurately reflects the values and struggles of unmanned aviation, squashes inaccuracies and misperceptions about the RPA community within the media, society, and military, values the community's significant contributions to national defense, and recognizes that killing remotely has unique challenges and unique problems. The secretive nature of the work does not help contribute to a positive culture but, as I have demonstrated throughout, you don't need to know classified details of the work to understand what a community endures.

Acceptance of this relatively new weapon will come with time. RPAs most certainly are not going away anytime soon. In fact, if the first couple of decades are any indication of future use, the RPA force will only continue to grow and evolve. Advocacy at the senior leader level will help to accelerate acceptance within the military and improve the overall culture. Dispersing RPA personnel during a nonoperational tour to other important jobs within the military will also help improve the culture as RPA disciples proselytize the benefits, advantages, and challenges to those outside the community.

Improving the culture starts with outsiders desiring to understand the RPA community and having factual knowledge about it. Throughout this book, I have endeavored to provide the most accurate information available concerning the RPA community. It is now up to future generations to determine if they desire to understand RPAs.

Chapter Four

Where Do We Go from Here?: Future Killer Robots

Now I am become Death, destroyer of worlds.
—Manhattan Project lead scientist Robert Oppenheimer quoting
the Bhagavad Gita upon witnessing the detonation of the first atomic
bomb on July 16, 1945

The Need for International Regulation

February 3, 2016, seemed like a day of RPA operations just like any other. An RPA was loaded with ordnance at the airfield and took off in search of targets, eventually finding what was identified as an ammunition storehouse for some insurgent fighters. The RPA crew launched a missile into the storehouse and a large explosion followed. Mission success! The strike ensured the enemy fighters would not be using *that* ammunition against friendly forces. As it happens, the enemy fighters in this instance were ISIS affiliates from Boko Haram and the friendly forces were the Nigerian Air Force flying a Chinese-made armed RPA. While this scenario may seem an anomaly, the use of Chinese-made armed RPAs is actually a growing trend throughout Africa and the Middle East.

In 1987 the U.S. and thirty-four other nations of the Missile Technology Control Regime (MTCR) agreed to informal regulations regarding which nations to export missiles to. The MTCR originated as a way of preventing the proliferation of weapons of mass destruction. However, once armed RPAs were developed, they were lumped under the same informal regulatory arms-control agreements in the MTCR and U.S. Conventional Arms Transfer (CAT) policies. So, in the last two decades, the U.S. has abided by these formal and informal regulations, exporting armed RPAs only to its allies in NATO. This situation left a huge gap in non-NATO countries that desired to have RPAs. And China, which is not a member of the MTCR, was there to fill the void.

This development is troubling for several reasons. Michael Horowitz, in "A Way to Rein in Drone Proliferation," touched on one major issue:

> This spread of military drones...is an important international security issue for several reasons. Chief among them is that many countries buy armed drones from exporters who care little about how the buyers use them. When the United States sells a weapon system to a partner or ally, it requires the buyer to abide not only by international law, but also by additional U.S. restrictions. The United States has leverage because it can cut off the supply of replacement parts and ammunition. When China sells weapons, it places no such restrictions. As a result, the buyer can use the weapon system without any concern for international laws and regulations. This makes it more likely that countries will use armed drones in ways that contribute to instability.

So armed Chinese RPAs are proliferating throughout Africa and the Middle East and, as a result, at the end of 2019 eighteen countries possessed armed RPAs while another dozen countries were seeking to own them. Regarding those countries, China shows little reluctance to sell armed RPAs to them and even less concern about how they use them. To make matters worse, the Pentagon says China can produce almost forty-two thousand drones by 2023.

China gains influence, access, and a strategic geographical foothold selling weapons to U.S. allies. To counter all these worrisome developments, the U.S. eased up its Conventional Arms Transfer Policy in April 2018 and will try to regain some influence with allies by selling them armed RPAs on a case-by-case basis. And that is how an arms race begins. But armed RPAs are only the beginning. Taken to its logical conclusion, armed RPAs will succumb to lethal autonomous weapons.

All It Takes Is One

In the very near future, negative psychological responses to killing with RPAs may be the least of our concerns compared to the complex problems surfacing with RPAs on a global scale. We are approaching three very important decision points regarding what the international community will allow with RPAs. The first two challenges, as discussed, are the lack of international proliferation regulations relating to armed RPAs and the lack of regulation of RPAs generally. The third challenge is the convergence of autonomy in robotics, artificial intelligence, machine learning, and lethal autonomous weapons, specifically in light of the first two challenges. We may not be able to get the RPA genie back in the bottle, but it is imperative that we act on this third challenge. In a report released in 2017, Harvard University's Belfer Center warned that lethal autonomous weapons could become as disruptive as nuclear weapons.

All it takes is one country to set off this lethal autonomous arms race. The other nations with the ability to do so will respond in kind, lest they run the risk of fielding an obsolete war machine that can't compete with the rapid decision-making and employment of their enemies' weapons. U.S. Air Force Colonel Jon Boyd referred to this process as an OODA—for "observe, orient, decide, and act"—loop. Boyd attributed U.S. fighter pilots' air-to-air kill success in the Korean War not to their superior jets but to their ability to act inside their opponents' decision cycle. They could observe, orient, decide, and act faster than their enemy.

What human can process information faster than a machine? Humans can no longer beat machines at games such as chess or *Jeopardy!*. Why would we think that humans could make decisions faster than lethal autonomous weapons? The human becomes the limiting factor in this process of making the decision to kill or not kill. Brief periods of delay or indecision by a human could be the difference between victory and defeat when facing an opponent who isn't hampered with a human in the decision-making loop. It's the difference between a Major League Baseball player deciding if a ninety-five mile-per-hour pitch is a ball or strike in .25 seconds and a Little Leaguer trying to do the same. It's competing on a different level. And all it takes is one. Who will that one be?

Well, no country wants to finish second in war. So all countries with the ability to move toward this objective are doing so. In the U.S., a bipartisan commission established by the 2019 National Defense Authorization Act and led by former Google CEO Eric Schmidt concluded that:

> The US military should adopt artificial intelligence urgently without letting debates over ethics and human control "paralyze AI development." In light of the choices being made by our strategic competitors, the United States must also examine AI through a military lens, *including concepts for AI-enabled autonomous operations.*

Commercial advances in robotics, urban air mobility (think Uber Air), self-driving cars, artificial intelligence/machine learning, and edge computing may actually be providing the critical research and development for the defense industries. 5G networks will provide critical infrastructure support required to employ autonomous robots. 5G networks can host more users and can transmit data at a faster rate, all things necessary to employ a lethal autonomous weapon. Indications are that lethal autonomous weapons will one day soon proliferate the battlefield.

For the Sake of Humanity, Humans Must Remain the Killers

I started this chapter with Oppenheimer's quote upon witnessing a destructive power without compare in human history. Oppenheimer understood that with unprecedented destructive power comes unprecedented responsibility. With the rapid advancement of autonomy in robotics, artificial intelligence, and machine learning, we are not very far from the decision points of unprecedented responsibility. Do we allow lethal autonomous weapons? Almost everyone I interviewed said that humans must remain in the decision-making loop to kill.

My intention in this book has been to show the humanity of the warriors undertaking this current fight and the effects that killing remotely with robots has on their lives. It would be simple, and perhaps the next logical step for some, to use these challenges as justification to state that a portion of those warriors killing with robots are suffering and the solution should be to remove the human from the loop and allow the robots to do the dirty business of war for us.

This is certainly not my intent and I do not advocate it. War is a human endeavor and there must be a human cost to undertaking it. For the sake of humanity, we must not think that any human life is so insignificant that we allow robots to make the decision to take the autonomous action to kill a human. It must always be difficult to kill another human being, lest we risk losing what makes us human to begin with.

Acknowledgments

I charged into this project with the unearned confidence on an aspiring author, not fully appreciating the time and effort required to research, write, and publish a book. I, like most new authors undoubtedly, underestimated the time required to complete the process. Every aspect of the project was new, exciting, and fresh mostly because I had no idea what I was doing. Thankfully I had a few seasoned mentors and a fantastic support team at home and at Little, Brown and Company, who provided guidance every step of the way and turned my vision—of telling the story of the challenges remote warriors face—into a reality.

As I neared retirement from twenty years of service in the Marine Corps, I started to get the itch to write a book about the unique and misunderstood community of warriors I belonged to for the last six years of my career. I had an idea and a title, and I cold pitched them to Lt. Col. Dave Grossman (U.S. Army, retired), the author of *On Killing* and *On Combat,* and widely considered an expert on the human response to killing. I fully expected the idea to die on the vine, but six days later I received a message from Dave that said he was intrigued and wanted to explore the idea further. At that point I was absolutely the dog surprised he caught the car. Now what? Dave advised me to conduct a thorough literature review and to interview as many people as I could before I started writing. Solid advice.

Off I went with my marching orders. I created an online survey and encouraged people to take it and provide their point of contact information if they were interested in participating in an interview. Two hundred and fifty-four people responded in total and more than fifty of them agreed to an interview. I am forever grateful to all who responded and took a chance that the survey was legitimate and not some ISIS phishing attempt.

Justin Rhode (Captain, USAF) saw my survey online and contacted me and said that he wanted to assist making sense of all the data associated with the surveys. I had met Justin in the RPA training pipeline in Pueblo, Colorado, and San Antonio, Texas, in 2014. We had been classmates, and I knew that his insight and expertise would be critical to getting this project right. It helped that Justin was a psychology major and an MQ-9 Reaper pilot in the U.S. Air Force as well. Justin pitched that he wanted to be a junior research assistant, but he became so much more throughout the research and writing. He evolved into a sounding board and a developmental editor for every chapter of the book. His contributions and fingerprints are sprinkled throughout this work. To this day he remains a close friend and we are working together to write the logical successor to *On Killing Remotely,* about lethal autonomous weapons.

The interviews I conducted were extremely insightful and revealed challenges that I never imagined. They were also cathartic for most who participated. Perhaps it was shortsighted on my part, but I never anticipated that people would open up to me so much about their struggles that linger from what their nation has asked them to do. Dave Grossman understood this from the thousands of conversations he conducted over decades of research and writing. Dave was right. You can never conduct too many interviews.

Some interviewees chose to remain anonymous, and although there are stories throughout the book without attribution, each one was told to me by a brave warrior who thought it important enough to spend a couple of hours on a Zoom call with me. To those who remained anonymous, thank you for your courage and assistance with this work.

Some interviewees allowed me to use their name or rank and first name or their full name. You have read about these heroes throughout, such as Douglas Wood, Aaron Garman, Christopher Herr, Tim Stack, Joshua Brooks, Eric Cooper, Alexander "Epic" MachPhail, Madhur

"Miso" Sawhaney. To these warriors, thank you for sharing your time and experiences, not only with me but with the readers. The human element of RPA work cannot be overstated.

David Funkhouser, a legend in Marine unmanned aviation, shared countless stories and also introduced me to the warriors of VC-6, who flew RQ-2 Pioneers during Desert Storm. Dale Bohannon, David Place, Randy McDonald, Ted Ferriter, Mark Ballinger, and Mark Rindler set the record straight on the first time in recorded history that an enemy surrendered to a robot. Thank you, gentlemen, and a job well done to VC-6 and RPV Companies 1 and 3 for paving the way.

I had the privilege of interviewing Scott Swanson (USAF, retired), the very first Predator pilot to kill in combat with an armed RPA. Scott told me to go read Richard Whittle's book *Predator* and circle back with him. Once again, solid advice. Scott also reviewed the manuscript before publication and has been a huge advocate throughout the entire process and I cannot thank him enough.

My good friend John Griffin (Lt. Col. USMC, retired, and associate professor at the Naval War College) wrote a few paragraphs for the book. John was my neighbor in California for a brief period and we served in Afghanistan at the same time in 2010. John is the kind of humble servant warrior we all aspire to be, and I learn something from him each time we talk. I sincerely appreciate his contributions to *On Killing Remotely*.

Once the manuscript was complete, I sent it to the legendary literary agent Richard Curtis. Richard plowed through the manuscript in a few days and told me this compelling story must be published. There are moments in life when you meet the right person at the right time. Richard was that person for me. His sage wisdom and guidance at every step of the process has turned *On Killing Remotely* into a reality, and I am forever grateful to him for his mentorship.

As word of this project spread I was introduced to experts who have thought deeply on the topic of remote killing and have dedicated diligent efforts to help these warriors. Dr. Peter Lee, the author of *Reaper Force,* Dave Blair (USAF Reaper pilot), and Karen House, a licensed counselor who was embedded with Air Force Reaper units were three people who greatly influenced my thinking. Thank you, Pete, Dave, and Karen, for all that you do for the RPA community.

Once Little, Brown and Company picked up *On Killing Remotely,* I was introduced to my editor, Terry Adams. Terry has become a friend and mentor throughout this process. He always had time for my dumb questions and was yet another person whose diligent efforts made this possible. Terry challenged me to be a better writer and to tell a story that is easily digestible and understandable. His patience and guidance were inspiring.

My production editor, Michael Noon, and copyeditor, Scott Wilson, really polished this work and made it what it is today. Their the best detail-oriented professionals I could have asked for to go through the book with a fine-tooth comb to catch my grammar and word-usage errors! (Just kidding about the improper use of "their," Mike and Scott.) But, seriously, they're the best!

Gabriel Leporati, Bowen Dunnan, and the entire publicity and marketing team at Little, Brown are amazing. They couldn't have made this process any easier for a newbie.

The Phantoms of VMU-3 were my original inspiration for this book. They showed me what teamwork, tenacity, and a fighting spirit could accomplish. To those I have served with, I thank you for your sacrifices and dedication to your trade. You are our nation's greatest treasure.

To everyone who believed in this idea and offered advice and encouragement along the way, please know that it inspired me to complete this daunting task.

Lastly, my support team at home enabled me to chase the dream of writing this book. My wife, Julie, my son, Hunter, and my daughter, Kai, didn't know what they signed up for when I said I wanted to write a book. Thank you for your patience, encouragement, love, and support throughout the long nights I spent tucked away in my office, banging away on the computer. You continue to be the light that guides my way.

Bibliography

Books

American Psychiatric Association. 2013. *Diagnostic and statistical manual of mental disorders* (5th ed). Washington, D.C.: APA.

Arkin, W. M. 2015. *Unmanned: Drones, data, and the illusion of perfect warfare.* New York: Little, Brown.

Asimov, I. 1950. "Runaround." *I, Robot* (The Isaac Asimov Collection ed.). New York: Doubleday.

Bessemer, H. 1905. *Sir Henry Bessemer, F.R.S.: an autobiography.* New York Public Library: Offices of "Engineering."

Chamayou, G. 2015. *Drone theory.* New York: The New Press.

Clausewitz, C. M. Von. 1976. *On war.* 2nd ed. Trans. M. Howard and P. Paret. Princeton, NJ: Princeton University Press.

Cockburn, A. 2015. *Kill chain: drones and the rise of high-tech assassins.* New York: Verso.

Du Picq, A. 1947. *Battle studies: ancient and modern battle.* Harrisburg, Pa.: Military Service Publishing Company.

Dyer, G. 1985. *War.* London: Guild Publishing.

Friedman, T. 2007. *The world is flat* (2nd revised and expanded ed.). New York: Farrar, Straus, and Giroux.

Gabriel, R. and K. Metz. 1992. *A short history of war: the evolution of warfare and weapons.* Strategic Studies Institute: U.S. Army War College, Carlisle Barracks, PA.

Gavin, J. M. 1958. *War and peace in the space age.* New York: Harper and Brothers.

Groom, W. 2013. *The aviators: Eddie Rickenbacker, Jimmy Doolittle, Charles Lindbergh, and the epic age of flight.* National Geographic Society.

Grossman, D. 2009. *On killing: the psychological cost of learning to kill in war and society,* revised ed. New York: Back Bay Books/Little, Brown and Company.

Grossman, D. and L. Christensen. 2008. *On combat: the psychology and physiology of deadly conflict in war and in peace.* 3rd ed. Millstadt, Illinois: Warrior Science Publications.

Holmes, R. 1986. *Acts of war: the behavior of men in battle.* New York: Free Press.

Johnson, P. 1983. *Modern times: the world from the twenties to the nineties.* New York: Harper and Row.

Keegan, J. 1993. *A history of warfare*. New York: Vintage Books.

Keegan, J. 2009. *The face of battle*. 11th ed. New York: Penguin.

Lee, P. 2018. *Reaper force: inside Britain's drone wars*. London: John Blake Publishing.

Lorenz, K. 1963. *On aggression*. New York: Bantam Books.

Marshall, S. L. A. 2000. *Men against fire: the problem of battle command*. Norman: University of Oklahoma Press.

Musharraf, P. 2006. *In the line of fire: a memoir*. New York: Free Press.

Nadelson, T. 2005. *Trained to kill: soldiers at war*. Baltimore, Maryland: The Johns Hopkins University Press.

Newman, A. 1997. *Follow me: the human element in leadership*. Novato, Calif.: Presidio.

O'Connell, R. 1990. *Of arms and men: a history of war, weapons, and aggression*. New York: Oxford University Press.

Pressfield, S. 2011. *The warrior ethos*. New York: Black Irish Entertainment.

Riza, M. S. 2013. *Killing without heart: limits on robotic warfare in an age of persistent conflict*. Washington, D.C.: Potomac Books.

Scahill, J. 2016. *The assassination complex*. New York: Simon & Schuster.

Shalit, B. 1988. *The psychology of conflict and combat*. New York: Praeger.

Solis, G. D. 2010. *The law of armed conflict: international humanitarian law in war*. New York: Cambridge University Press.

Troxel, W., R. A. Shih, E. Pedersen, L. Geyer, M. P. Fisher, B. A. Griffin, A. C. Haas, J. R. Kurz, and P. S. Steingberg. 2015. *Sleep in the military: promoting healthy sleep among U.S. servicemembers*. Santa Monica, Calif.: Rand.

Velicovich, B. and C. S. Stewart. 2017. *Drone warrior*. New York: HarperCollins.

Whittle, R. 2014. *Predator: the secret origins of the drone revolution*. New York: Picador.

Articles

Armour, C. and J. Ross. 2017. The health and well-being of military drone operators and intelligence analysts: a systematic review. *Military Psychology* 29 (2): 83–98.

Bandura, A. Feb. 2017. Disengaging morality from robotic warfare. *The Psychologist*. https://thepsychologist.bps.org.uk/volume-30/february-2017/disengaging-morality-robotic-war

Barclay, N. Oct. 9, 2015. Top 10 RPA myths debunked. Creech Air Force Base, Nev. creech.af.mil/News/Article-Display/Article/669932/busted-top-10-rpa-myths-debunked/

Barron, L. G., T. R. Caretta, and M. R. Rose. March 2016. Aptitude and trait predictors of manned and unmanned aircraft pilot job performance. *Military Psychology* 28 (2): 65–77.

Belenky, G., D. M. Penetar, D. Thorne, K. Popp, J. Leu, M. Thomas, H. Sing, T. Balkin, N. Wesensten, and D. Redmond. 1994. The effects of sleep deprivation on performance during continuous combat operations. In: Marriott, B., ed. *Food components to enhance performance*. Washington, D.C.: National Academy Press,127–35.

Blair, D. and K. House. Nov. 12, 2017. Avengers in wrath: moral agency and trauma prevention for remote warriors. lawfareblog.com/avengers-wrath-moral-agency-andtrauma-prevention-remote-warriors.

Boyne, W. J. July 2009. How the predator grew teeth. *AIR FORCE Magazine*.

Campo, J. 2015. Distance in war: the experience of MQ-1 and MQ-9 aircrew. *Air and Space Power Journal.*

Campo, J. 2015. From a distance: the psychology of killing with remotely piloted aircraft. PhD dissertation. Maxwell Air Force Base, Montgomery, Alabama.

Cannon, C. May 25, 2017. RPA culture continues innovation. USAF News. af.mil /News/Article- Display/Article/1192440/rpa- culture-continues-innovation/

Castano, E., B. Leidner, and P. Slawuta. 2008. Social identification processes, group dynamics and the behavior of combatants. *International Review of the Red Cross,* 90(870): 259–271.

Chapa, J. O. 2017. Remotely piloted aircraft, risk, and killing as sacrifice: the cost of remote warfare. *Journal of Military Ethics.*

Chatterjee, P. Mar. 6, 2015. A chilling new post-traumatic stress disorder: why drone pilots are quitting in record numbers. Salon.com. salon.com/2015/03/06/a_chill ing_new_post_traumatic_stress_disorder_why_drone_pilots_are_quitting_in_record _numbers_partner/

Chappelle, W., K. McDonald, W. Thompson, and J. Swearengen, Dec. 2012. Prevalence of high emotional distress and symptoms of post-traumatic stress disorder in U.S. air force active duty remotely piloted aircraft operators (2010 USAFSAM Survey Results). School of Aerospace Medicine Wright Patterson AFB OH, Aerospace Medicine Department.

Chappelle, W., T. Goodman, L. Reardon, and W. Thompson. Jun. 2014. An analysis of post-traumatic stress symptoms in United States air force drone operators. *Journal of Anxiety Disorders* 28(5).

Chappelle, W., T. Goodman, K. McDonald, L. Prince, B. Ray-Sannerud, and W. Thompson. Aug. 2014. Symptoms of psychological distress and post-traumatic stress disorder in United States air force "drone" operators. *Military Medicine.*

Chappelle, W., K. McDonald, T. Goodman, L. Prince, B. Ray-Sannerud, and W. Thompson. Sep 2014. Assessment of occupational burnout in United States air force predator/reaper "drone" operators. *Military Psychology* 26(5–6).

Chappelle, W., S. Cowper, T. Goodman, K. McDonald, L. Prince, and W. Thompson. Mar. 2015. Reassessment of psychological distress and post-traumatic stress disorder in United States air force distributed common ground system operators. *Military Medicine* 180 (3, suppl.).

Chappelle, W. S. Cowper, T. Goodman, K. McDonald, L. Prince, and W. Thompson. Jun. 2015. Prevalence of posttraumatic stress symptoms in United States air force intelligence, surveillance, and reconnaissance agency imagery analysts. *Psychological Trauma Theory Research Practice and Policy* 8(1).

Chappelle, W., C. Bryan, T. Goodman, L. Prince, W. Thompson. Jun. 4, 2017. Occupational stressors, burnout, and predictors of suicide ideation among U.S. air force remote warriors. *Military Behavioral Health* 6(1).

Chappelle, W., T. Goodman, L. Prince, E. Skinner, and J. Swearingen. Mar 29, 2018. Emotional reactions to killing in remotely piloted aircraft crewmembers during and following weapon strikes. *Military Behavioral Health* 6(4).

Chappelle, W., T. Goodman, L. Prince, and L. Reardon. March 2019. Combat and operational risk factors for post-traumatic stress disorder symptom criteria among United States air force remotely piloted aircraft "drone" warfighters. *Journal of Anxiety Disorders* (62).

Chow, D. Nov. 5, 2013. Drone wars: pilots reveal debilitating stress beyond virtual battlefield. *Live Science.*

Church, A. June 2011. Ramp up. *AIR FORCE Magazine.*

Conner, R. Mar. 9, 2018. The predator, a drone that transformed military combat. Smithsonian National Air and Space Museum.

Correll, J. T. March 2010. The emergence of smart bombs. *AIR FORCE Magazine.*

Cortright, D. Jan. 9, 2012. License to kill. cato-unbound.org/2012/01/09/david-cortright/license-kill.

Department of Defense Unmanned Aerial Vehicle (UAV) Roadmap. April 2001. globalsecurity.org/intell/library/reports/2001/uavr0401.htm#_Toc509971279DoD

DoD Unmanned Systems Integrated Roadmap. FY2011-2036, 2011. fas.org/irp/program/collect/usroadmap2011.pdf

DoD Unmanned Systems Integrated Roadmap 2017-2042.

Dowd, A. Dec. 15, 2016. Moral hazard: drones and the risk of risk-free war. Providencemag.com.

Drew, C. and D. Philipps. Jun. 16, 2015. As stress drives off drone operators, air force must cut flights. *New York Times.*

Drew, C. Mar. 16, 2009. Drones are weapons of choice in fighting Qaeda. *New York Times.*

Duray, J. Jan. 23, 2018. Forever deployed: why "combat-to-dwell" reform for MQ-9 crews is beyond overdue. Warontherocks.com.

Fiske, S. T. 2004. Social beings: core motives in social psychology. Hoboken, NJ: J. Wiley.

Gettinger, D. Apr. 21, 2014. Burdens of war: PTSD and drone crews. Center for the Study of the Drone.

Government Accountability Office Report 19-155 to Congress. February 2019. Unmanned aerial systems. Air Force pilot promotions rates have increased but oversight process of some positions could be enhanced.

Griffin, R. M. n.d. The health risks of shift work. WebMD.com.

Hennigen, W. J. Feb. 22, 2016. A fast growing club: countries that use drones for killing by remote control. *Los Angeles Times.*

Hijazi, A., C. Ferguson, H. Hall, M. Hovee, F. Ferraro, and S.Wilcox. 2017. Psychological dimensions of drone warfare. *Current Psychology* 1(12).

Hillman, K. n.d. Why do we identify with fictional characters? HealthGuidance.org.

Hitchens, T. Nov. 5, 2019. DoD should consider truly autonomous weapons: bipartisan AI commission. Breakingdefense.com.

Horowitz, M. C. Nov. 30, 2018. A way to rein in drone proliferation. Thebulletin.org.

Khan, M. I. Feb. 11, 2016. Pakistan's secretive Houbara bustard hunting industry. BBC News.

Kirkpatrick, J. Jan. 26, 2016. Military drone operators risk a serious injury, and it's one you might not expect. Slate.com.

Kreuzer, M. P. Sep.–Oct. 2015. Examining future of unmanned combat aerial vehicles and remotely piloted aircraft—analysis. *Air and Space Power Journal,* 29(5).

Latson, J. Oct. 21, 2014. How Edison invented the lightbulb—and lots of myths about himself. *Time.*

Lee, P. 2012. Remoteness, risk and aircrew ethos. *Air Power Review* 15(1): 1–20.

Lifton, R. J. Jul. 1, 2013. The dimensions of contemporary war and violence: how to reclaim humanity from a continuing revolution in the technology of killing. *Bulletin of the Atomic Scientists.*

Martin, R. Dec. 18, 2011. Report: high levels of burnout in U.S. drone pilots. *Morning Edition,* National Public Radio.

McGrath, T. Sep. 16, 2014. The US is now involved in 134 wars or none, depending on your definition of "war." PRI.org.

McLeod, S. 2017. The Milgram Experiment. Simplypsychology.org.

Michel, A. H. Sep. 4, 2015. Drones in popular culture. Center for the Study of the Drone at Bard College.

Murray, K. F. Oct. 2016. Marine aviation readiness: solving the problem. *Marine Corps Gazette.*

Otto, J. L. and B. J. Webber. March 2013. Mental health diagnoses and counseling among pilots of remotely piloted aircraft in the United States air force. *MSMR* 20(3).

Pinchevski, A. 2016. Screen trauma: visual media and post-traumatic stress disorder. *Theory, Culture, and Society* 33(4): 55–75.

Purkiss, J. and J. Serle. Jan. 17, 2017. Obama's covert drone war in numbers: ten times more strikes than Bush. The Bureau of Investigative Journalism.

Ricks, T. E., and D. J. Morris. Jun. 2, 2015. Can drone operators get PTSD? *Foreign Policy.*

Roblin, S. Sep. 29, 2019. Chinese drones are going to war all over the Middle East and Africa: and killing lots of people. NationalInterest.org.

Seligman, L. May 11, 2016. Air force pilots, maintainers on F-35 pros and cons. *Defense News.*

Stauton, J. Nov. 30, 2015. Does losing one sense improve the others? *Science ABC.*

Thompson, J. Aug. 30, 2018. Phantom of Takur Ghar: the predator above Roberts Ridge. Creech Air Force Base.

Turak, N. Feb. 21, 2019. Pentagon is scrambling as China "sells the hell out of" armed drones to US allies. CNBC.com.

The United States Army in Afghanistan, Operation EnduringFreedom: October 2001–March 2002. history.army.mil/brochures/Afghanistan/Operation Enduring Freedom.htm

Walker, L. May 15, 2015. What "good kill" gets wrong about drone warfare. *Newsweek.*

Wilson, G. I. Sep. 28, 2011. The psychology of killer drones—actions against our foes, reactions affecting us.

Websites

skybrary.aero/index.php/Cockpit_Automation__Advantages_and_Safety_Challenges af.mil/About-Us/Fact-Sheets/Display/Article/104525/air-force-distributed common-ground-system/history.navy.mil/research/archives/command operations-reports/aviation-commands/vc-composite-squadrons/vc-6.html. VC-6 Command Chronology 1991.

thebureauinvestigates.com/stories/2017-01-01/drone-wars-the-full-data cancen-tral.com/can-stats/history-of-the-can

Unattributed. Oct. 5, 2016. Drone visuality—the psychology of killing. Cybersecurity-intelligence.com Posted October 5, 2016 cybersecurityintelligence.com/blog/drone-visuality-the-psychology-of-killing-1726.html

sleepfoundation.org/articles/what-circadian-rhythm

Cluff, J. R. USAF. Biography. af.mil/About-Us/Biographies/Display/Article/1211793/brigadier-general-james-r-cluff/

nbcdfw.com/news/local/Officers-Who-Used-Robot-to-Kill-Ambush-Sniper-Wont
    Face-Charges-Grand-Jury-471979213.html
belfercenter.org/sites/default/files/files/publication/AI%20NatSec%20-%20final.pdf
newamerica.org/in-depth/world-of-drones/4-who-has-what-countries-developing
    armed-drones/

Poem

John Gillespie Magee Jr. in his sonnet "High Flight."

Radio Show Interview

Shapiro, A. Nov. 12, 2018. "The Cleaners" Looks at Who Cleans Up The Internet's
    Toxic Content. NPR. *All Things Considered.* npr.org/2018/11/12/667118322/
    the-cleaners-looks-at-who-cleans-up-the-internets-toxic-content

Award Citations

Medal of Honor Sgt. Dakota Meyer. United States Marine Corps. Awarded 9/15/2011
    for actions on 9/8/2009, Kunar Province, Afghanistan. cmohs.org/recipient-detail/
    3480/meyer-dakota.php

Index

Langley, Samuel, 19–20
last resort, 160, 161–162, 188
latency, 84
Latson, Jennifer, 222
launch and recovery element (LRE), 84–86, 90, 312, xi, xiii
The Law of Armed Conflict: International Humanitarian law in War (Solis), 167
Law of Armed Conflict (LOAC), 173, 174
learner role, in Milgram experiment, 229
Lee, Peter, 179, 181, 266
Leidner, Bernhard, 247
LeMay, Curtis, 161
lethal miniature aerial munitions (LMAMs), 38–39
lethargy, 218
Libya, 164, 176
"License to Kill" (Cortright), 162
Lifton, Robert Jay, 94
limber, 9
Lindbergh, Charles, 13
line of sight (LOS), 104
Little Boy (bomb), 14
long shifts, dangers of, 220
long-range bombers, 202–203
Lorenz, Konrad, 247
low-intensity conflicts, 157

Mabus, Ray, 76
machine guns, 11–12, 12
machine learning, 319–320
MacPhail, Alexander, 234–235
Magee, John Gillespie, Jr., 12
Magnavox, 276
maintainers, 174
major combat operations (MCO), 80
Mamelukes, 18
man-in-the-loop, 155
Maori warriors, 8
March of Dimes, 211
marginalization, 30–32
Marine Air Group (MAG), 297
"Marine Aviation Readiness" (Murray), 8
Marine Corps, 29
    Base Quantico, xiv
    insignia, 291–292
    rifle range score, 184
    training programs, 53–54

Marine Special Operations Command (MARSOC), 38, 40, 45–46
Marine Unmanned Aerial Vehicle Squadron, 297
marksman badge, 185
Marr, Benard, 68
Marshall, S.L.A., 248
Mattis, James, 123, 304
max range, 202–203
Maxim, Hiram, 11–12
McChrystal, Stanley, 169
meat eaters, 296–298
mechanical distance, 192–193, 207, 208
Medal of Honor, 28, 169, 171
medals, 299–300
mediated trauma, 116–119
medium altitude long endurance (MALE), 44–47
melatonin, 222
memory gap, 133–134
Men Against Fire (Marshall), 248
mental armor, 217–218
mental health diagnoses, 91–92
mental health professionals, 100
mental preparation, 142–143
Metz, Karen, 5, 8
Meyer, Dakota L., 169–171, 182
Middle East, 318
MiG-25, 77
Milgram, Stanley, 228–230, 239
Milgram's experiment, 228–230, 239
military aged male (MAM), 187
military culture, 289–303, 291
    in aviation, 294–296
    changing, 316
    definitions of, 290
    flight suits, 291
    fratricide, 294–296
    insignia, 291–292
    meat eaters, 296–298
    medals, 299–300
    media portrayal of RPA personnel, 301–302
    opportunities for advancement, 301
    overview, 289–291
    pecking order, 296–297
    poor, 289–290
    promotions, 298–299
    subcultures, 290, 296
    wings, 291–292

"Military Drone Operators Risk a Serious Injury" (Kirkpatrick), 179
military jargons, 187
military occupational specialty (MOS), 290
Military Psychology, 294
Minié, Claude-Étienne, 10
Minié ball, 10
Missile Technology Control Regime (MTCR), 318
mission control element (MCE), 84–86, xii
missions, 68–80
    authority figures as audiences, 227–228
    hunting high-value individuals, 195–199
    intelligence, surveillance, and reconnaissance (ISR), 70
    killer, 72–75
    pattern of life, 71–72, 135, 146, 194
    security clearance, 101–102
    tasking, 78–80
missions of, connection to, 63
Mitchell, Billy, 161
mob mentality, 247–248
Modern Times (Johnson), 163
moral distance, 207, 208–209
"Moral Hazard: Drones and the Risks of Risk-Free War" (Dowd), 162
moral injury, 190
morality, 178
Morgenthau, Hans, 151
Mormons, 242
MQ-1 Predator, 20, 25–27, 29, 44, 130, 265, 279, 280–281, 285–286
MQ-1C Gray Eagle, 44–45, 59, 139, 260, 265, 285
MQ-25, 54
MQ-25A Stingray, 76
MQ-4 Triton, 47
MQ-8 Fire Scout, 44
MQ-9 Reaper, 20, 47–48, 50, 53, 77–78, 103, 121, 130, 139, 235, 265, 280–281, 284, 285, 285–286, xvi
murder hole, 43
murder of children, 264, 265
Murray, Kevin, 80
mutual surveillance, 248
My Lai massacre, 248

Nadelson, Theodore, 205–206
Nagorno-Karabakh conflict, 39

Nangarhar Province, Afghanistan, 263
Napoleon, Prince, 9
Napoleon III, Emperor, 9–10
napping, 224
nation building, 291
National Defense Authorization Act, 320
National Geospatial-Intelligence Agency, 32
National Public Radio, 91
National Sleep Foundation, 221
Naval Air Station Pensacola, 279
naval aviators, 292
Naval flight officer (NFO), 292
Naval War College, 158
Navy
    RPA training program, 54
    Stingray, 76
negative reinforcement, 298
negative responses, 143
Newman, Aubrey, 211
Niccol, Andrew, 301
nicknames, dehumanizing, 186–187
nicotine, 224
Niger, 164
Nigerian Air Force, 317
nine-line, 61
Nintendo medal, 300
Nintendo warriors, 302
nips, 186
Nobel Peace Prize, 151
non-state actors, 165, 177
nontraditional ISR (NT-ISR) mission, 80
North Atlantic Treaty Organisation (NATO), 163, 174, 318
North Vietnamese, 186
Northrop Grumman, 75–76
nuclear weapons, 13–14

Obama, Barack, 31, 151
obedience to authority, 228–230
offensive air support (OAS), 72
Omar, Mullah, xiv
On Combat (Grossman/Christensen), 7, 12, 52, 57, 63, 133, 220, 307–309
On Killing, 56–57, 64, 67, 112, 120–121, 191, 193, 196, 202, 207, 231, 237, 249, 298
On War (Clausewitz), 179, 202
onager, 8
one-sided killing, 171

About the Author

Lt. Col. Wayne Phelps (USMC, Retired) served five deployments in Iraq and Afghanistan between 2001 and 2012. His military career coincided with the escalating use of drones as weapons, and in the two years prior to his retirement in 2018 he served as commanding officer of a squadron that deployed four Unmanned Aircraft System teams abroad to fight violent extremist organizations. He lives in Austin, Texas.